ELECTRONICS

D1148456

ELECTRONICS

Malcolm Plant

Hodder & Stoughton

A MEMBER OF THE HODDER HEADLINE GROUP

British Library Cataloguing in Publication Data

Plant, M. (Malcolm), 1936–
  Teach yourself electronics
  1. Electronics
  speaking Students
  I. Title
  537.5
  ISBN 0 340 42230 0

First published 1988
Reissued 1992
Impression number 15 14 13 12 11 10 9
Year   1999 1998 1997 1996

Typeset by Rowland Phototypesetting Ltd, Bury St Edmunds, Suffolk.
Printed in Great Britain for Hodder & Stoughton Educational, a division of Hodder Headline Plc, 338 Euston Road, London NW1 3BH by Cox & Wyman Ltd, Reading, Berks.

Contents

Introduction

Not chaos-like, together crushed and bruised,
But, as the world harmoniously confused:
Where order in variety we see,
And where, though all things differ, all agree.
(Alexander Pope '*Windsor Forest*')

For new students and enthusiasts, electronics must seem a confusing jungle of loosely related concepts and objects with strange names. But Pope made an observation about the flora in Windsor Forest which could help us unravel the complexities of electronics. An amateur botanist (as Pope surely was!) might start by recognising the similarities, rather than the differences, between the component parts of a forest. Thus, at first glance, an oak tree bears little resemblance to a bramble. But both have branches, leaves and flowers, and both need sunlight and water to survive. Similarly, a television looks quite different from a radio-controlled model aircraft. But both have aerials, amplifiers and switches, and both need electrical power and a transmitter to be of any use. This is the 'order in variety' recognised by Pope.

This book emphasises the purpose and function of circuit building blocks such as switches, rectifiers, timers, amplifiers, oscillators, logic gates and counters that, in various combinations, make up the many electronic systems we use. Chapter 1 describes some of these uses, such as medicine, leisure and space research, and briefly traces the more significant milestones in the history of electronics. The following four chapters progressively develop an understanding of electronics, starting with the concepts of current flow in simple circuits and the electrical properties of semiconductors (Chapters 2 to 4). Chapter 5 emphasises the importance of Ohm's law and the use of resistors in simple circuits.

The purpose of Chapters 6 to 11 is to explain how these basic concepts determine the function of a number of important circuit building blocks. Thus Chapter 6 shows that timers, used in all kinds of ways from washing machines to cameras, depend on the properties of capacitors. The way diodes are used in power supplies is described in Chapter 7. The properties of transistors and their function in audio amplifiers is outlined in Chapter 8. The use of logic gates in decision-making circuits is described in Chapter 9. Chapter 10 shows the use of flip-flops in electronic counters. And the way light-emitting diodes and liquid crystal devices are used to produce electronic displays is discussed in Chapter 11.

The fascinating field of microelectronics is the focus of Chapter 13. It includes the use of flip-flops in computer memories, the unique properties of gallium arsenide (a rival to silicon), and the manufacture of silicon chips. The way electronic building blocks are used in control, instrumentation and telecommunications systems is described in the remaining chapters. Finally, a summary of the jargon of electronics is included at the end of the book.

1

Electronics Today and Yesterday

1.1 The Electronic Age

Electronics is about using things such as transistors and silicon chips
to make electricity work for us. The transistor was invented in the
late 1940s followed by silicon chips in the early 1960s. A silicon chip
(or integrated circuit) may contain up to several thousand transistors and other devices, all formed on a sliver of silicon so small you
could lose it under your finger nail. Miniaturising electronic circuits
in this way is called **microelectronics** and it has had far-reaching
effects on nearly all aspects of life. Microelectronics is influencing
the way we store, process and distribute information; it is changing the way we design and manufacture industrial goods; it is im-
proving the diagnosis of illnesses; and it is affecting financial and
business affairs, and a whole range of social, educational and

Fig. 1.1 A silicon chip

political activities. Nowadays, we take for granted the way electronics makes our lives more comfortable, enjoyable and exciting, so let's begin by looking at some of its benefits.

1.2 Consumer goods

We are surrounded by products in and around our homes which make use of electronics in one way or another. Washing machines, burglar alarms and toasters have electronic parts in them, not forgetting hi-fi systems, videos and microcomputers. For example, the quality and durability of recorded music has been greatly improved by the compact disk (CD). This is a plastic disk about 120 mm in diameter and 1.2 mm thick, which stores sound in the form of microscopically small pits along a track that spirals out from the centre of the disk. Each pit is about a thousandth of a millimetre long and about a tenth as deep. The track is so narrow that thirty tracks are about as wide as a human hair. Sixty minutes of sound recording requires about ten million such pits. The compact disk is coated with a layer of reflecting aluminium over which is placed a protective film of transparent plastic.

Fig. 1.2   A portable compact disk player
Courtesy: Morphy Richards CE Ltd

The disk is rotated at high speed in a player in which a finely focused laser beam 'reads" the information on the disk. The stream of on–off pulses of laser light is converted to electrical signals which are processed to produce sound that is almost free from distortion (e.g. surface hiss) which can mar the sound produced by a conventional record player using a vinyl disk. A similar technique for recording TV pictures was invented by Philips. It is called LaserVision and the first commercial video player was introduced in the USA in 1978. LaserVision uses a larger (300 mm) disk to store about an hour's worth of picture information as a succession of pits along a tight spiral track which takes 54 000 turns to cover the disk's surface; there are 480 tracks per millimetre.

There has been a revolution in photography, too, brought about by the use of miniaturised circuits built into a camera. Sensors measure the light received through the camera lens, and an integrated circuit acts as a tiny computer which works out the precise exposure time a picture requires and sets the shutter speed accordingly. Automatic focusing and a liquid crystal digital display that shows exposure time, film speed, number of pictures taken, programme selection, and so on, are increasingly common facilities on today's cameras.

Fig. 1.3 A modern electronically controlled camera
Courtesy: Ferranti

1.3 Communications electronics

Many man-made satellites are put into orbit round the Earth to provide world-wide communications by radio and television. Countries that can afford to make the powerful rockets necessary to launch communications satellites can profitably launch satellites for other nations. Europe has such a launch rocket, called Ariane, which can put the giant Intelsat-5 satellites into orbit. Each satellite is capable of handling 12 000 telephone calls and two colour TV channels simultaneously. Communications satellites that are designed to broadcast TV programmes direct to homes are usually put into geostationary orbit. This is an orbit which keeps pace with the Earth's rotation so the satellite remains in a fixed position above a particular place on the Earth. Homes making use of these direct broadcast satellites (DBS) require only a small indoor dish antenna pointed at the satellite that 'hangs' in the sky.

To keep in touch with interplanetary spacecraft that operate on or near other planets requires very large dish antennae on Earth and complex electronic equipment on board the spacecraft. For example, the *Voyager 2* spacecraft which is now touring the remote regions of the solar system has sent back intriguing information

Fig. 1.4   Olympus, a new generation of direct broadcast TV satellites
Courtesy: British Aerospace

Fig. 1.5 Giotto, the spacecraft that flew through Halley's Comet
Courtesy: British Aerospace

about the planets Jupiter, Saturn and Uranus. Contact with *Voyager 2* is expected to continue until it has passed Neptune in 1989. But it will take this spacecraft 350 000 years to reach another star! Some spacecraft have been instructed to send back pictures from the surfaces of the planets Mars and Venus. Others have flown through the cloud of dust and gas in the tails of comets. As Halley's comet swept round the Sun in 1986, it was examined at close quarters by the *Giotto* spacecraft. Protected by a shield from possible damage by high-speed dust particles, *Giotto* used its dish antenna to send back the first colour pictures ever taken of this famous comet before it departed to distant parts of the solar system for another 76 years.

In modern communications systems, hair-thin glass fibres are being used instead of conventional wires. Enormous amounts of information are carried down these fibres, not by electricity but by pulses of laser light reflected repeatedly from the inside wall of the fibre. The combined use of optics and electronics in this way is called **optoelectronics**. There are good reasons for developing optical communications systems based on sending pulses of laser light along glass fibres. Strong magnetic fields, e.g. from lightning and electrical machinery, do not interfere with the messages carried on the laser beam, and broken fibres are not a fire hazard since any light which escapes is harmless. Also glass fibres are cheaper and lighter than copper wires, and a single glass fibre can carry far more information than a copper wire. Optical fibres are particularly useful for replacing copper wires on aircraft and ships since fewer connections are required and they weigh less than copper wires.

1.4 Computer electronics

Compared with the first valve computers of the 1940s, today's computers show how dramatic the advances in electronics have been. Power-hungry, room-sized and unreliable, these early computers have been replaced by a variety of compact and efficient computers such as calculators, microcomputers and mainframe computers.

Such computers, like so many other products and systems, owe their efficiency and compactness to the development of integrated circuits on silicon chips. This is a small piece of silicon on which complex miniaturised circuits are made by photographic and chemical processes. A microprocessor, one or more of which is the 'brain' of a microcomputer, is the most complex silicon chip made today. Such chips are 'housed' in a plastics package which has two parallel rows of pins for connecting it to other devices to make it into a working computer. Thus the ubiquitous microcomputer has a memory (another integrated circuit) for storing information, a keyboard (or other input device) for entering information into this memory, a power supply, and a visual display unit (VDU) to display information.

Computers have to work fast if they are to be of any use in ballistic missiles and weather centres. But making fast work of calculations is not the only useful characteristic of a computer. For example, because it can be programmed through its keyboard, or from an external store of instructions on magnetic tape or disk, a microcomputer can be used for playing computer games, filing and accounting, diagnosing illnesses, controlling a home heating system, and accessing databases for business and travel information via the telephone. These uses for a microcomputer are enhanced by a wide range of 'add-ons' such as printers, digitisers, speech processors and synthesisers.

1.5 Control electronics

Electronics is an essential part of modern control systems. For example, anti-skid braking systems on some cars depend on electronic devices that ensure wheels do not lock in a skid. Safety in passenger and military aircraft relies on complex control systems. And the regulation of temperature and pressure in chemical factories and nuclear power stations depends on electronic control systems for the safe and efficient operation of the plant.

In the home, the power of electric drills and foodmixers can be controlled at the turn of a knob. In the greenhouse, automatic control of temperature and humidity is possible. And under the control of a program, the microprocessor in a washing machine instructs control devices (e.g. motors, heaters and valves) to carry out a pre-arranged sequence of washing activities.

Robots are computer-controlled machines. They are used for tasks which are repetitive (and often boring for humans) such as welding and painting, and mounting of windows in a car manufacturing plant. In such places as nuclear reactors and interplanetary space, computer-controlled robots can work where it would be hazardous for humans to do so.

1.6 Medical electronics

Electronic equipment is widely used to diagnose the cause of illnesses, and to help surgeons operate. For example, the electrocardiograph (ECG) is an electronic instrument which can help in diagnosing heart defects. By means of electrodes attached to the body, the ECG picks up and analyses the electrical signals generated by the heart. The ECG is just one of a number of instruments available to nurses and doctors in hospitals. A complete patient-monitoring system would also record body temperature, blood pressure and heart rate and warn staff should there be any change.

The ultrasonic scanner is a computerised medical instrument for providing images of the body's internal organs. It uses high-frequency sound waves generated by an ultrasonic device held in contact with the body. The echo pattern of the ultrasonic waves bouncing from inside the body is processed by a computer to give an image on a VDU. It is believed that these ultrasonic waves are safer than X-rays for examining the developing baby in the womb.

Bioelectronics is to do with implanting electronic devices in the body to help correct defects and so help people lead fuller lives. Now that microelectronics devices are so small and require very little electrical power to operate them, there are new challenges for bioelectronics. For example, research is being directed towards developing a video camera small enough to implant in a blind person's eye socket. Connected to the nerves of the visual cortex, the camera could enable a blind person to see.

A long-term goal of medicine is the manufacture of a biochip – a silicon chip that is implanted in the body to make an on-the-spot check of chemical activity. For example, a biochip designed to respond to adrenalin in the blood could operate alongside a heart pacemaker. The pulse of the pacemaker could then adapt to the varying degrees of excitement as the normal heart does. But what if a biochip could be implanted in the brain – a sort of 'bolt-on intelligence'?

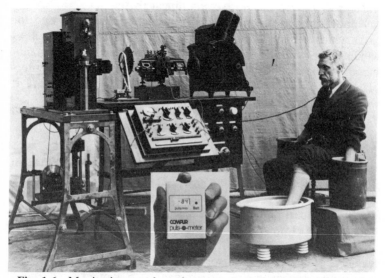

Fig. 1.6   Monitoring your heart beat wasn't much fun in the old days!
Courtesy: Picker International Ltd
The inset shows a modern way of doing it.
Courtesy: Centronic Health Care Services

1.7 Looking back

Less than a hundred years ago, electronics was unknown. There were no radios, televisions, computers, robots or artificial satellites, none of the products and services described above and which we now take for granted. In such a short time there has been a revolution in the ways we communicate, control, measure, take care of people and enjoy ourselves. This revolution came naturally out of a study of electricity, an old science not unknown to the Greeks over 2000 years ago. Electricity was a subject of great interest to Victorian scientists, and to Sir William Crookes in particular.

1.8 The discovery of cathode rays

The beginnings of electronics can be traced back to the discovery of cathode rays in the closing years of the last century. These mysterious rays had been seen when an electrical discharge took place between two electrodes in a glass tube from which most of the air had been removed. Sir William Crookes called these rays 'cathode rays' since they seemed to start at the negative electrode (the cathode) and moved towards the positive electrode (the anode).

At that time nobody had any idea what cathode rays really were. But during a historic lecture at the Royal Institution in London in April 1897, Sir J. J. Thomson declared that cathode rays were actually small, rapidly moving electrical charges. Later these charges were called electrons after the Greek word for amber.

Amber is fossilised resin from trees and has peculiar properties, as the ancient Greeks had found. If rubbed with fur or a dry cloth, it has the power to attract small pieces of dust and fluff. Neither the Greeks, nor the scientists who devoted so much time to studying its properties in the period from the 17th century, had a successful explanation of why amber behaved in this way. But the discovery of the electron provided the answer.

We now know that the electrical behaviour of amber (and of many other electrical insulators) is caused by static electricity. The friction between the cloth and amber causes electrons to be transferred from the cloth to the amber where they stay put to give amber an overall negative charge. This negative charge causes the amber to attract small bits of material to it. The vinyl disks we play on a record turntable show this effect, too.

1.9 The invention of the valve

The first practical application of cathode rays was the invention of the **thermionic valve** by Sir John A. Fleming in 1904. In this device, electrons were produced by heating a wire (the filament) in an evacuated glass bulb. The word 'thermionic' comes from 'therm' meaning 'heat', and 'ion' meaning 'charged particle', i.e. the electron. In a valve, negatively charged electrons ejected from the heated filament (the cathode) moved rapidly to a more positive anode. The flow of electrons stopped if the anode became more negative than the cathode. This electronic component was called a diode since it had two electrodes. And it acted like a valve because electrons flowed through it only in one direction, from the cathode to the anode, not in the opposite direction.

It did not take long for an American, Lee de Forest, to make a much more interesting and useful thermionic valve. By adding a third electrode made of a mesh of fine wire through which the electrons could pass, he produced a triode. By adjusting the voltage on this third electrode (called the grid), he was able to make the triode behave like a switch and, more important, as an amplifier of weak signals. The triode made it possible to communicate over long distances by radio, and this development was demonstrated dramatically in 1912 when the luxury liner *Titanic* collided with an iceberg in the Atlantic Ocean. As this 'unsinkable' liner was going down, her radio operator broadcast an SOS radio signal using Morse Code which was picked up by ships in the area.

1.10 The beginnings of radio and TV

Strangely, the First World War (1914–18) did little to stimulate applications for thermionic valves. But immediately after the war,

Fig. 1.7 A 1920s Echodyne superhet radio receiver
Courtesy: The Science Museum

the demands of entertainment gave electronics a push which has gained strength ever since. In London the British Broadcasting Corporation was formed, and in 1922 its transmitter (call sign 2LO) went on the air. Firms such as Marconi, HMV and Echo made radio sets from components and valves supplied by Mazda, Ozram, Brimar and others.

The second major boost to the emerging electronics industry was the start of regular television transmissions from Alexandra Palace in London in 1936. But at that time the public had little interest in television, which was hardly surprising as the pictures produced by John Logie Baird's mechanical scanning system were not very clear. By the time EMI had developed an electronic scanning system which gave much better pictures, the Second World War had begun. The Alexandra Palace transmitter was closed down abruptly in September 1939 at the end of a Mickey Mouse film. Britain feared that Germany might use the transmission as a homing beacon for its aircraft to bomb London.

1.11 Radar and the Second World War

However, from 1939 to 1945 there were important advances in electronics. Perhaps the most significant invention was radar, developed in Britain to locate enemy aircraft and ships.

The word radar is an acronym, for it is formed from the words RAdio Detection And Range-finding. Radar was made possible by the invention of a high-power thermionic valve called the magnetron. This device produced high frequency pulses of radio energy which were reflected back from aircraft or ships to reveal their range and bearing. Magnetrons are the source of microwaves in microwave cookers.

1.12 The invention of the transistor

In the period immediately following the Second World War, there was a major step forward in electronics brought about by the invention of the first working transistor. In 1948 Shockley and his co-workers Barden and Brattain, working in the Bell Telephone Laboratories in the USA, demonstrated that a transistor could amplify electrical signals and act like a switch. Actually, the idea of using the element germanium to produce a solid-state equivalent of the valve had been worked out about 25 years earlier. However, the way electricity moved in semiconductors, as these germanium-based devices were called, was not well understood. Furthermore, until the 1950s it was not possible to produce germanium with the high purity required to make useful transistors.

These transistors turned out to be successful rivals to the thermionic valve. They were cheaper to make since their manufacture could be automated. They were smaller, more rugged and had a longer life than valves, and they required less electrical power to work them. Once silicon began to replace germanium as the basic semiconductor for making transistors in the 1960s, it was clear that the valve could never compete with the transistor for reliability and compactness.

For consider ENIAC (another acronym, for Electronic Numerical Integrator And Calculator). This was a computer which used valves and was built at the University of Pennsylvania in the 1940s. ENIAC filled a room, was worked by 18 000 valves, needed 200 kW of electrical power to work it, had a mass of 30 tonnes and cost a million dollars. The transistorised desk-top calculator of the 1960s was battery-powered, had a mass of a few kilograms and was capable of far more sophisticated calculations than ENIAC was. And this trend towards low-cost yet more complex functions, to greater reliability yet lower power consumption, continues to be an important characteristic of developments in electronics. ENIAC is now regarded as a first-generation computer, and the transistorised computers which followed it in the 1960s as second-generation computers.

1.13 Silicon chips make an impact

The first integrated circuits were made during the early 1960s. Techniques were developed for forming up to a few hundred transistors on a silicon chip and linking them together to produce a working circuit. Computers using this new technology are now known as third-generation computers. The Apollo spacecraft which took man to the Moon in the late 1960s and early 1970s used these computers for navigation and control. The stimulus to miniaturise circuits in this way came from three main areas: weapons technology, the 'space race' and commercial activity.

Modern weapons systems depend for their success on circuits which are small, light, quick to respond, absolutely reliable, and which use hardly any electrical power. Miniature circuits on silicon chips offer these advantages. The 'space race' began when Russia

Fig. 1.8 The 'three ages' of electronics: valve, transistor and integrated
circuits

launched Sputnik in 1957. At first America's response was unsatis-
factory, but she gained ground during the 1960s and Americans
walked on the Moon by the end of the decade. Lacking the
enormously powerful booster rockets developed by Russia,
America needed compact and complex spacecraft which stimulated
the design of small and reliable control, communications and
computer equipment. During the 1970s, spin-off from military
interests and the space race stimulated the growth of an electronics
industry bent first on creating electronics goods and then satisfying
the demand for them in the home, the office and industry.

During the 1970s, the number of transistors integrated on a
silicon chip doubled every year and this trend continues. Along with
this increasing circuit complexity has been a similar doubling in the
information processing power of the silicon chip. The most important type of chip is the microprocessor. It contains most of the
components needed to operate as the central processor unit of a
computer. It is a highly complex device which can be programmed
to do a variety of tasks. This versatility means that it acts as the
'brain' in a wide variety of devices. These fourth-generation computers have become faster and cheaper; they are now used in
industrial robots and sewing machines, in space stations and

toasters, in medical equipment and computer games. Their programmability and cheapness are their strength. The microprocessor brings the story to the present day.

2

The Basis of Electronics

2.1 Electrons in atoms

Section 1.8 explained that electronics began with the discovery that cathode rays were actually beams of negatively charged particles called electrons. Nowadays, cathode rays are widely used to 'draw' information on television and radar screens. In these devices, electrons are temporarily 'free' as they move through the vacuum inside the television or radar tube. But electrons are normally constituents of atoms.

Atoms are extremely small 'bits' of material – millions of them lie side by side across the diameter of the dot at the end of this sentence. All atoms have a nucleus. Fig. 2.1 shows a simple model of an atom in which atomic particles called protons and neutrons have their home in the nucleus. Electrons, however, make up an 'electron cloud' outside the nucleus. The nucleus is very small compared with the overall size of an atom – say the size of an orange compared with the vast volume of an English cathedral. Using this model of an atom, you can imagine the electrons to be flies in the

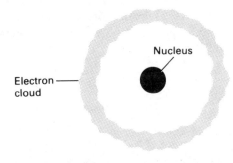

Fig. 2.1 A simple model of an atom

cathedral. Most of the mass of an atom lies in the nucleus. In fact, the neutron and proton have about equal masses whereas the electron has a mass about 2000 times smaller than either particle.

From the point of view of electronics, the two most important properties of an electron are its electrical charge and its small mass. Its electrical charge means that it can be moved by an electric field, as in a telephone wire. Its small mass means that the path of a beam of electrons can be rapidly bent as in a television tube. An electron carries a negative charge and the proton an equal positive charge. Since these charges are opposite, protons and electrons attract each other. It is this attraction that keeps electrons in the electron cloud surrounding the nucleus of an atom. The neutron carries no electrical charge, i.e. it is neutral, and it does not have any part to play in making electrons stay in the electronic cloud.

2.2 Atomic structure

Hydrogen and oxygen are two very common substances since their atoms go to make up that very useful liquid called water. Fig. 2.2 shows that a hydrogen atom has the simplest structure of all atoms, since it has just one proton in its nucleus and one electron in the space surrounding the nucleus. It is this single proton that tells us the atom is hydrogen since it is the number of protons in the nucleus of an atom which determines the physical and chemical properties of that atom. The electrical charge on a proton is equal and opposite to the charge on an electron, making the normal hydrogen atom electrically neutral.

An oxygen atom has a more complicated structure as shown in Fig. 2.3. Its nucleus is made up of eight protons and eight neutrons. Thus an electrically neutral oxygen atom has eight electrons in the space surrounding the nucleus. There are more than 100 different atoms in the universe, all made from the three main fundamental particles, neutrons, protons and electrons. The table opposite summarises the atomic structure of just a few of these atoms.

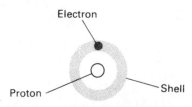

Fig. 2.2   A hydrogen atom

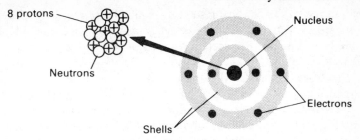

Fig. 2.3 An oxygen atom

Neutrons combine with protons to make up the nuclei of all atoms, but they do not carry an electrical charge. In spite of the neutron's zero charge, neutrons and protons do attract each other strongly when they are close together in the nucleus of an atom. When this force is overcome, as is the case when uranium atoms are 'split' in nuclear reactors and atomic bombs, an enormous amount of energy is released. However, electrons are held much more weakly to atoms, and it is on this weakness that electronics is based.

Atom	Number of protons in nucleus	Number of neutrons in nucleus	Number of electrons in shells
Hydrogen	1	0	1
Oxygen	8	8	8
Copper	29	34	29
Silver	47	61	47
Silicon	14	14	14
Germanium	32	40	32
Carbon	6	6	6
Iron	26	30	26

Some information about atoms

2.3 Conductors, insulators and semiconductors

The reason why some materials, such as copper, are good electrical conductors is that they contain 'free' electrons which are quite weakly bound to the nuclei of the atoms of the material. These electrons can be moved easily by connecting the material across a battery. Copper and aluminium are good electrical conductors and are used in electronics to allow electrons to flow easily between one device and another. Electrons are more strongly attracted to their parent nuclei in electrical insulators, which therefore do not have any free electrons. Thus electrical insulators such as glass, polythene and mica are used to resist the flow of electrons between electronic devices.

Electronics is to do with the use of semiconductors as well as conductors and insulators. Semiconductors are the basis of electronics devices such as transistors and diodes, heat sensors and light emitters, integrated circuits and many other devices. As its name suggests, a semiconductor has an electrical resistance that falls somewhere between that of a conductor and that of an insulator.

Two of the commonest semiconductors are silicon and germanium. They are important in electronics because their resistance can be controlled to good effect. There are two ways of doing this. First there is the effect of heat on a semiconductor. At very low temperatures semiconductors happen to be good electrical insulators. But as their temperature increases they become increasingly better electrical conductors so that at everyday temperatures they allow some current to flow through them. Generally, this drop in resistance with temperature increase is a nuisance, although some devices, such as thermistors, do make use of this effect.

The second way of controlling the electrical resistance of silicon and germanium is to add minute amounts of carefully selected substances to them. We need to know something about the atomic structure of germanium and silicon to understand the effect this has.

2.4 Silicon atoms

Germanium is now rarely used in electronic devices, which are mostly based on silicon. Fig. 2.4 shows a model of a silicon atom that has fourteen electrons surrounding a nucleus containing fourteen

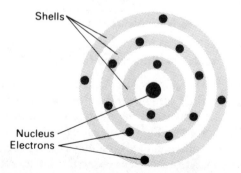

Fig. 2.4   A silicon atom

protons and fourteen neutrons. The part of this structure that makes silicon useful to electronics is the way the electrons are arranged in what are known as shells surrounding the nucleus. There are two electrons in the inner shell, eight in the next shell, and four in the outer shell. It is the four electrons in the outer shell, known as the valency shell, which make pure silicon a crystalline material.

In a crystal of pure silicon, each of the four outer electrons forms what is known as a covalent bond with an electron from a neighbouring silicon atom. Fig. 2.5 shows how the pairing of electrons uses up every one of these outer electrons. An orderly arrangement of silicon atoms results and gives pure silicon its crystalline structure. There are no free electrons available to make pure silicon conduct electricity and so it is an insulator. At least, it is an insulator at low temperatures, and a perfect insulator at the absolute zero of temperature ($-273\,°C$). But at everyday temperatures, silicon conducts electricity a little, not much but enough to make silicon a bit of a problem when it is used in transistors. However, we are not so much interested in how an increase of temperature reduces the resistance of silicon, but in what happens to its resistance when a small amount of an 'impurity' is added to it.

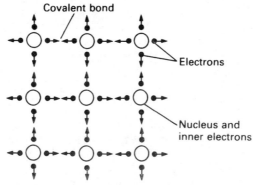

Fig. 2.5 The pairing of electrons in atoms

2.5 n-type and p-type semiconductors

Once a very pure crystal of silicon has been manufactured, the silicon is 'doped' with impurity atoms! These atoms are chosen so that they make a 'bad fit' in the crystal structure of silicon, due to the impurity atoms having too many or too few electrons in their outer shells. Depending on the impurity, two types of semiconductor are produced in this way, n-type or p-type.

An **n-type** semiconductor is produced by doping silicon with, for example, phosphorus. A phosphorus atom has five electrons in its outer shell. Fig. 2.6 shows what happens when an atom of phosphorus is embedded in the crystal structure of pure silicon. Four of the five outer phosphorus electrons form covalent bonds with neighbouring silicon atoms, leaving a fifth unpaired electron. This unattached electron is now weakly bound to its parent phosphorus atom and it is therefore free to wander about. Phosphorus is said to be a *donor* impurity since each atom of phosphorus can donate (give away) an electron. The addition of phosphorus has therefore changed the electrical properties of silicon. It has become an electrical conductor due to the presence of free electrons donated by phosphorus atoms. An n-type semiconductor has been produced ('n' for negative).

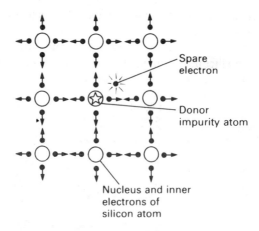

Spare electron

Donor impurity atom

Nucleus and inner electrons of silicon atom

Fig. 2.6 How a donor atom produces free electrons

A **p-type** semiconductor is produced by doping silicon with atoms such as boron which have three electrons in their outer shells. Fig. 2.7 shows what happens when a boron atom becomes embedded in the crystal structure of silicon. Three of its outer electrons become paired with neighbouring silicon atoms, leaving one unpaired silicon electron. This electron is not available for conduction but it will accept another electron to pair with it. The vacancy created in silicon by doping it with boron is known (not surprisingly!) as a 'hole'. Since this hole attracts an electron, it behaves as if it had a positive charge. Boron is said to be an *acceptor* impurity since

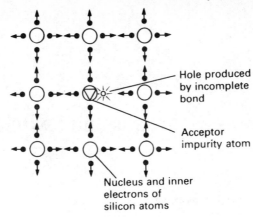

Fig. 2.7 How an acceptor atom produces a hole

its atoms can accept an electron from other nearby atoms. The presence of holes which act as positive charges in boron-doped silicon produces a p-type semiconductor ('p' for positive).

3

Simple Circuits and Switches

3.1 Making electrons move

Just as a mechanical force is required to make a snooker ball move on a snooker table, an electrical force is required to make an electron move in a conductor. If the conductor is a wire as shown in Fig. 3.1, electrons move through it if there is a difference of electrical force between its ends. This force is called a **potential difference (p.d.)** (symbol V) and is measured in volts (symbol V). Since a lamp lights when it is connected across the terminals of a battery, there must be a potential difference between these terminals. In this case the potential difference is known as an electromotive force (e.m.f.) (symbol E) and this is also measured in volts.

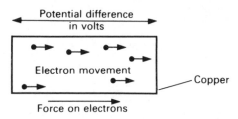

Fig. 3.1   The flow of electrons through copper

The flow of electrons through the lamp shown in Fig. 3.2 is called an electric **current** (symbol I) and is measured in amperes (symbol A). It is the flow of electrons through the conductor that constitutes the electric current. A great many electrons are on the move when a current of one ampere flows through a conductor. In fact, about six million, million, million move past a point in the circuit each second! Since the electrons carry an electric charge, the current is a flow of charge through the circuit. **Electrical charge** (symbol Q) is measured in coulombs (symbol C). When a current of one ampere

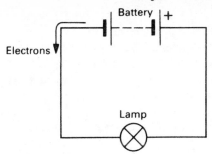

Fig. 3.2 A simple circuit

flows through a circuit, the rate at which charge flows is one coulomb per second. Thus

1 ampere = 1 coulomb per second

The flow of current in a good conductor such as copper is due to the movement of millions of free electrons which are directly available from copper atoms. But in an n-type semiconductor, current flow is due to the movement of weakly held free electrons donated by donor impurity atoms as shown in Fig. 3.3. The free electrons present in an n-type semiconductor due to the presence of donor atoms are called majority charge carriers. There are also a very few electrons and holes produced by the effect of heat, which breaks covalent bonds between silicon atoms. The holes in an n-type semiconductor are called minority charge carriers. The contribution of minority charge carriers to the current that flows in

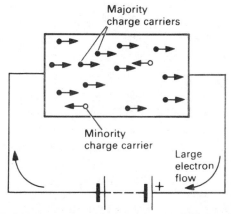

Fig. 3.3 The flow of electrons in an n-type semiconductor

n-type semiconductors is negligible, but note that the holes flow in the opposite direction to the electrons.

Fig. 3.4 shows how current flows through a p-type semiconductor in which holes are created in the crystal structure of silicon by acceptor impurity atoms. As for an n-type semiconductor, electrons are once again the mobile charges that make up the current flowing through the material. But they move through the material by alternately filling and vacating these holes. In fact, current flow through a p-type semiconductor is most conveniently considered to be due to the movement of positively charged holes. The flow of holes can be likened to the movement of an empty seat in a row of theatre seats. If a seat at the end of a row becomes vacant (a hole), the person next to it might move into it and so produce a vacant seat. If the next person moved into the newly created vacant seat (and so on), an empty seat would appear to move along the row. In the case of a p-type semiconductor, the people correspond to the electrons and the empty seat to the hole. Note that a small number of electrons and holes is produced by the effect of heat on silicon atoms. In a p-type semiconductor, the electrons are known as minority charge carriers. They flow in the opposite direction to the holes and make a negligible contribution to the current.

The coulomb (unit of electric charge) and the ampere (unit of electric current) honour the names of two French physicists. Charles Augustin de Coulomb, born in 1736, was a military engineer in his younger days, but his penchant for scientific experiment led him to formulate the inverse square law of forces between

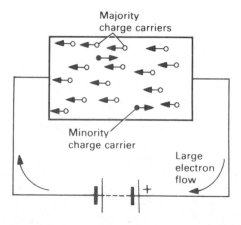

Fig. 3.4   The flow of holes in a p-type semiconductor

electrically charged spheres. André Marie Ampère, born in 1775, mastered advanced mathematics by the age of twelve and became something of an absent-minded professor – he once forgot to keep an invitation to dine with the Emperor Napoleon! Ampère studied the magnetic force that surrounds a current-carrying conductor and he is responsible for the idea that current flows from the positive terminal of a battery, which he believed to have more 'electrical fluid' than the negative terminal.

3.2 Series and parallel circuits

Devices such as lamps, switches and batteries are known as **components**. They are the individual items which are connected together to make a **circuit**. Fig. 3.5 shows a simple circuit in which a battery B_1, which has an e.m.f. of 6 V, makes a current I flow through lamps L_1 and L_2. The switch SW_1 has two positions, open and closed. When the switch is closed, it offers a low resistance and allows current to flow round the circuit. If SW_1 is open, it offers a high resistance (the resistance of air) and stops current flowing through the circuit. The current comprises electrons which flow from the negative to the positive terminal of the battery. Before electrons were discovered, it was thought that current flows from the positive to the negative terminal of a battery. This direction is known as **conventional current** and is usually marked on circuit diagrams. The circuit of Fig. 3.5 is called a **series circuit** since the battery, lamps and switch are connected one after the other. In a series circuit, the current flowing is the same at any point in the circuit so that the same current flows through each lamp. A room light which is switched on by a wall light switch is connected in series with the mains supply.

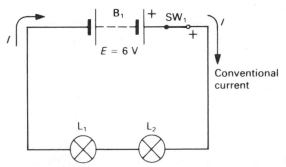

Fig. 3.5 Two lamps connected in series

Another common arrangement of components is shown in Fig. 3.6. In this circuit, two identical lamps L_1 and L_2 are connected side-by-side to a 6 V battery B_1. Each lamp has the same p.d. across it, i.e. 6 V, since each is connected across the battery. With this p.d., the current flowing through each lamp is 0.06 A. Thus 0.06 A flows through each lamp when both switches are closed. So the total current provided by the battery is 0.12 A. The circuit of Fig. 3.6 is known as a **parallel circuit**. In this circuit, switch SW_1 independently controls lamp L_1, and SW_2 independently controls L_2. In the home, different appliances are connected in parallel with the mains supply so that they can be individually controlled by their respective on/off switches.

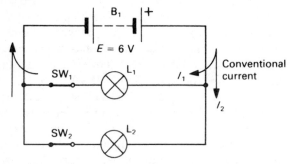

Fig. 3.6　Two lamps connected in parallel

3.3　Resistance and Ohm's law

The relative ease with which electrons flow through a conductor is measured by its electrical **resistance** (symbol R). Thus copper has a low resistance since it is a good conductor of electricity; glass has a high resistance since it is a very poor conductor, i.e. an insulator. On the other hand, silicon has a resistance that falls somewhere between conductors and insulators and is known as a semiconductor. The electrical resistance of a conductor is measured in units of ohms and is defined by the following equation:

$$\text{resistance} = \frac{\text{p.d. across the conductor}}{\text{current through the conductor}}$$

or
$$R = \frac{V}{I}$$

This equation can be used to find the resistance of one of the filament lamps show in Fig. 3.6. The p.d. across each lamp is 6 V, which is the e.m.f. of the battery. The current through each lamp is 0.06 A. Thus the resistance of the lamp is found as follows:

$$R = \frac{6\,V}{0.06\,A} = 100\text{ ohms}$$

The unit of electrical resistance is the ohm, whose symbol is the Greek letter omega (Ω). So the lamp has resistance 100 Ω.

Of course, we could use the equation to find the current through a device, or the p.d. across a device. We would then need to use alternative forms of the above equation. Fig. 3.7 helps to get the equations right.

To find R, cover R and $R = \dfrac{V}{I}$

To find V, cover V and $V = I \times R$

To find I, cover I and $I = \dfrac{V}{R}$

For example, suppose you want to find the current flowing through a 12 V car headlamp bulb which has resistance of 3 Ω.

Since the current $I = V/R$,

$$I = \frac{12\,V}{3} = 4\,A.$$

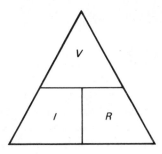

Fig. 3.7 A triangle for working out V, I and R

If the resistance of a device is constant for a range of different values of potential difference and current, the device is said to be linear or ohmic, and it obeys Ohm's law. The graph in Fig. 3.8a shows the behaviour of an ohmic device for which the ratio V/I is constant. In electronics, devices called resistors obey Ohm's law quite closely. If the device is non-ohmic then V/I is not a constant and it does not obey Ohm's law as the graph in Fig. 3.8b shows. This famous law, and the unit of resistance, honour the only major contribution to electricity made by the German physicist George Simon Ohm in 1827.

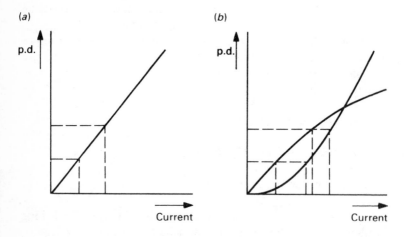

Fig. 3.8 Comparison of ohmic and non-ohmic conductors. (a) Ohmic: if the p.d. doubles, the current doubles; (b) Non-ohmic: if the p.d. doubles the current is more than or less than doubled

3.4 Types of switch

Switches are used to switch current on or off in a circuit. A switch is 'on', i.e. closed, if it allows current to pass through it, and it is 'off', i.e. open, if it stops current flowing through it. Simple on–off switches are in use dozens of times a day. A car's ignition, heater and radio are operated using on–off switches. So are cookers, hi-fi systems, television, radio and burglar alarms in the home. And the keyboard or keypad switches of calculators, computers and electronic games are simple on–off switches.

Many switches in everyday use require a mechanical force to operate them. The force brings together, or separates, electrically conducting metal contacts. Three types of mechanical switch are shown in Fig. 3.9. The **push-button switch** (Fig. 3.9a) is a simple push-to-make, release-to-break type. There are two circuit symbols for this type of switch depending on whether pushing 'makes' or 'breaks' the contacts. **Slide** and **toggle** switches (Fig. 3.9b) are

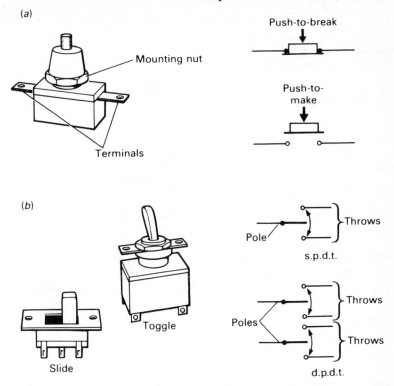

Fig. 3.9 (a) Push-to-make, release-to-break switch;; (b) slide and toggle switches

generally made either as single-pole, double-throw (s.p.d.t.), or as double-pole, double-throw (d.p.d.t). The poles of these switches are the number of separate circuits the switch will make or break simultaneously. Thus a d.p.d.t switch can operate two separate circuits at the same time. A s.p.d.t switch is sometimes known as a change-over switch, since the pair of contacts which is made changes over as the switch is operated.

The microswitch shown in Fig. 3.10a is simply a sensitive mechanical switch. It is usually fitted with a lever so that only a small force is required to operate it. This force causes contacts in the switch to open and close. The rotary switch shown in Fig. 3.10b has one or more fixed contacts (its poles) which makes contact with moveable contacts mounted on its spindle. Thus a number of switching combinations can be made: e.g. 1-pole, 4-way; 2-pole, 6-way; 4-pole, 3-way and 6-pole, 2-way.

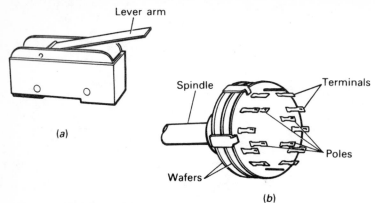

Fig. 3.10   (a) Microswitch, and (b) a rotary switch

The two switches shown in Fig. 3.11 are magnetically operated. The **reed switch** (Fig. 3.11a) has two flat and flexible ferromagnetic reeds which are easily magnetised and demagnetised. The contacts are protected by sealing them inside a glass envelope, usually containing nitrogen (or some other chemically inert gas) to reduce corrosion of the contacts. If a permanent magnet is brought close to a reed switch, the reeds become temporarily magnetised, attract each other and make electrical contact, so closing a circuit. On removing the magnet, the reeds lose their magnetism and separate and open a circuit. The reed switch is a *proximity* switch, since it is operated by the nearness of the magnet. And since the reeds are protected by the glass envelope, it is ideal for use in atmospheres containing explosive gases. Furthermore, it is a fast switch and can operate up to 2000 times per minute over a lifetime of more than a 1000 million switching operations.

The electromagnetic relay (Fig. 3.11b) has one or more pairs of contacts which are opened and closed by current flowing through an energising coil. The current has the effect of magnetising a soft-iron plate which is drawn to the coil and opens and closes the switch contacts at the end of the contact arms. The contacts are arranged as s.p.d.t or d.p.d.t, for example. The electromagnetic relay is a particularly useful switch for two reasons: the small current which energises it enables a much larger current to be switched via its contacts; and the energising current is completely isolated from the circuit which is switched on and off via its contacts.

In addition to the mechanical switches described above, electronics makes use of switches that have no moving parts. These are switches based on semiconductors, and include the transistor, which is widely used in control circuits and computer memories, the optoswitch, which operates when there is a change of light intensity, and triacs for the control of power supplied to mains-operated drills and foodmixers.

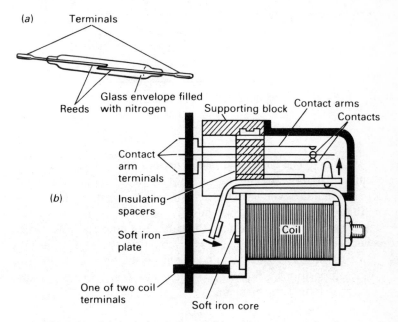

Fig. 3.11 (a) A reed switch, and (b) an electromagnetic relay

3.5 Uses for switches

This section describes how to use some of these mechanical switches.

(a) Door bell switch (Fig. 3.12)
This simple application of a single push-to-make, release-to-break switch (Fig. 3.9a) enables a door bell to be operated when the switch SW_1 is pressed.

(b) Two-way light switch (Fig. 3.13)
Two single-pole, double-throw (s.p.d.t) switches (Fig. 3.9b) are used in this circuit to enable a lamp to be switched on and off from two independent positions, e.g. from the top and bottom of a flight of stairs.

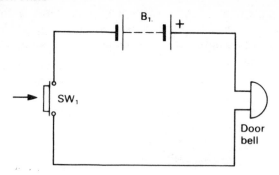

Fig. 3.12   A door bell switch circuit

CAUTION: The circuit uses a 6 V battery as a low-voltage power source. Do not attempt to wire up a similar circuit for mains operation unless you have the help of a skilled electrician.

In the position shown, switches SW_1 and SW_2 are in states which enable current to flow through the lamp. But note that SW_1 can be switched to state 2 which isolates the lamp from the battery. Or SW_2 can be switched to state 1 to do the same thing. And in either of these off positions, SW_1 may be switched to state 1, or SW_2 to state 2, to switch the lamp on again.

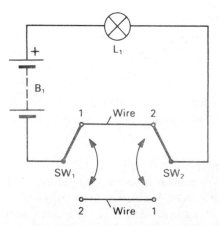

Fig. 3.13   A two-way switch circuit

(c) Motor reversing circuit (Fig. 3.14)

This uses one double-pole, double-throw toggle or slide switch (Fig. 3.9b) to reverse the direction of rotation of the motor. The motor must be a direct current (d.c.) type, i.e. one that reverses direction when the current flows in the opposite direction through it. The switch is shown in both of the two possible states. The dotted lines trace the path of the current from the battery via the switch contacts through the d.c. motor.

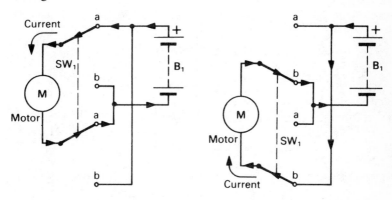

(a) Motor drives one way   (b) Motor drives opposite way

Fig. 3.14 A reversing circuit for an electric motor (the broken line indicates that the two parts of SW_1 operate together)

(d) Wheel counter (Fig. 3.15)

Here is a simple use of a microswitch (Fig. 3.10a) to count the number of revolutions made by a wheel. Each time the wheel rotates, a cam operates the arm on the microswitch. Since the microswitch has been fitted with a roller at the end of its lever arm, little frictional force is required to operate it.

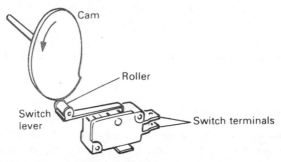

Fig. 3.15 A wheel counter

(e) Channel selector (Fig. 3.16)

A single-pole, 12-way rotary switch (Fig. 3.10b) can easily be used to select any one of 12 channels of information, e.g. radio stations 'piped' to hotel rooms and selected by a rotary switch by the side of the bed. Rotation of the switch connects one of the incoming channels to the loudspeaker.

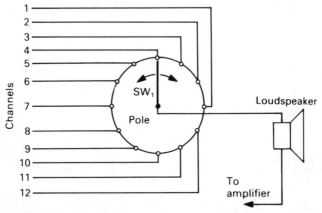

Fig. 3.16   A channel selector

(f) Burglar alarm (Fig. 3.17)

The reed switch (Fig. 3.11a) can be used as a proximity switch to detect when a door or window is opened. Two small magnets are required, one inset in the door (it can be made invisible), and the other next to the reed switch which is inset in the door frame. The latter magnet ensures that the reeds on this reed switch are closed when the door is open so that the alarm sounds if the 'alarm-set' switch is closed. However, when the door is shut, the magnetic fields of the two magnets cancel each other in the region of the reeds and the reeds are open. Hence the reeds close and the alarm sounds only when the door is opened. More than one door and window can be monitored in this way, merely by connecting all the reed switches in parallel with each other so that the closure of any reed switch sounds the alarm.

3.6 Simple digital circuits

Fig. 3.18a shows a simple circuit, comprising a switch SW_1 connected in series with a battery B_1 and a lamp L_1. When the switch is closed, the lamp is on; when it is opened, the lamp is off. Because

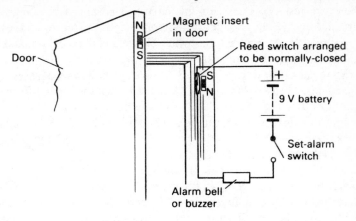

Fig. 3.17 A burglar alarm

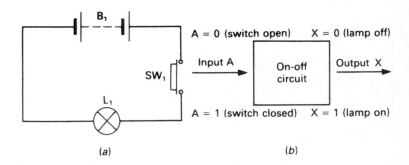

Fig. 3.18 (a) A simple on/off circuit; (b) the circuit shown as a functional black box; (c) the truth table for this on/off circuit

there are just two states for this circuit, it is said to be a digital circuit. If we use the two binary numbers, 1 and 0, to represent these two states, 1 can represent the switch closed and the lamp on, and 0 the switch open and lamp off.

The 'black box' drawn in Fig. 3.18b is a symbolic way of showing the circuit. The two states of the switch represent the input information (1 or 0) to the box. And the two states of the lamp represent the output information from the box. The table in Fig. 3.18c summarises the output and input information – it is called a **truth table** since it 'tells the truth' about the function of the circuit.

Now look at the more complicated circuit shown in Fig. 3.19a in which two switches, SW_1 and SW_2, are connected in series. Note that the lamp cannot light unless SW_1 *and* SW_2 are closed. As the black box in Fig. 3.19b shows, the switches provide input information and the lamp indicates the output information. The truth table in Fig. 3.19c summarises the function of this digital circuit which is known as an AND gate. It 'says' that the output state has a binary value of 1 (lamp on) only if switch SW_1 and switch SW_2 each have a value of binary 1 (both switches closed). If either or both switches are set to binary 0 (i.e. are open), the output is binary 0 (i.e. the

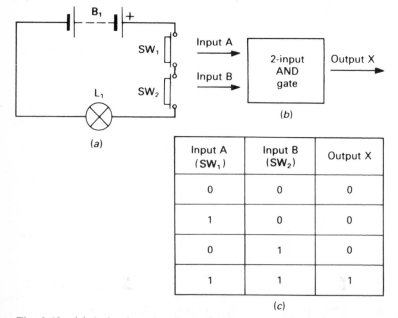

Input A (SW_1)	Input B (SW_2)	Output X
0	0	0
1	0	0
0	1	0
1	1	1

(c)

Fig. 3.19   (a) A simple series circuit; (b) the circuit shown as a functional black box; (c) the truth table for this 2-input AND gate

lamp is off). Note that this digital circuit is called a 'gate' because the switches open and close to control the information reaching the output.

A second simple digital circuit is shown in Fig. 3.20a in which the two switches are connected in parallel. In this circuit the lamp lights if switch SW_1 *or* switch SW_2 is closed. The lamp also lights if both switches are closed. This OR gate has the truth table shown in Fig. 3.20b which summarises the values of the output information for all values of the input information.

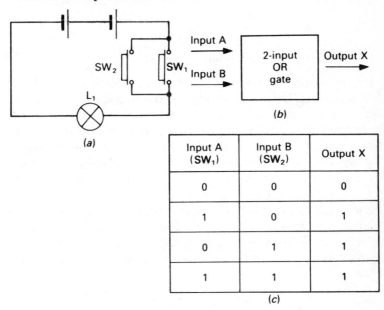

Input A (SW_1)	Input B (SW_2)	Output X
0	0	0
1	0	1
0	1	1
1	1	1

(c)

Fig. 3.20 (a) A simple parallel circuit; (b) the circuit shown as a functional black box; (c) the truth table for this 2-input OR gate

The branch of electronics introduced by the above series and parallel circuits is known as **digital logic**. The AND and OR gates are called **logic gates** since their output states are the logical (i.e. predictable) result of a certain combination of input states. These logic gates, and others besides, are of great importance to electronics nowadays for they are used in calculators, watches and computers. Many thousands of gates are built from transistors in the form of integrated circuits. This type of digital logic is discussed more fully in Chapter 9.

4

Signals and Systems

4.1 Direct current and alternating current

The type of current delivered by batteries in a torch is called **direct current** (d.c.). Direct current always flows in one direction; in this case it is conventional current that flows from the positive to the negative terminal of the battery. Though this d.c. current may vary in strength, it does not change direction. A steady direct current is shown in Fig. 4.1a. As time passes, the current remains at a steady level and in the same direction. If the source of current is a battery, the current will eventually fall to zero but not change direction. In a graph such as this, the plus (+) and minus (−) signs are used to indicate the two possible directions of current flow through the circuit. Fig. 4.1b shows a varying direct current, one that would be produced if a switch was regularly opened and closed in a simple

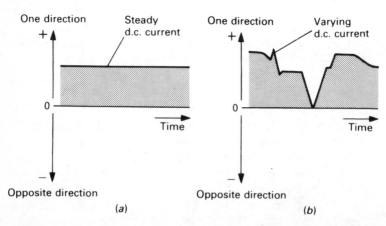

Fig. 4.1 Graphs of (a) a steady d.c. current, and (b) a varying d.c. current

circuit. Closing the switch causes the current to rise abruptly to a maximum value where it remains steady until the switch is opened, whereupon it falls to zero.

An **alternating current** (a.c.) is one that flows regularly first in one direction and then in the opposite direction. Fig. 4.2a shows the variation in the alternating current from the mains supply. It smoothly increases to a maximum in one direction and then falls to zero before increasing to a maximum in the opposite direction. For the mains supply, the regular variation in current is called a *sinusoidal* waveform since the graph is a sine wave. A *square wave* a.c. waveform is shown in Fig. 4.2b.

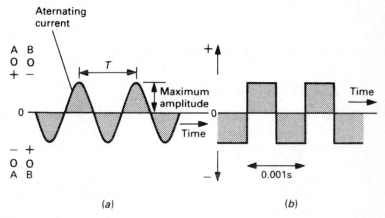

Fig. 4.2 Graphs of (a) a sinusoidal a.c. waveform, and (b) a square wave a.c. waveform

There are three characteristics of a regularly repeating waveform of the type shown in Fig. 4.2a. These are amplitude, period and frequency. The **amplitude** is the value of the current or voltage (or of any other quantity which varies in this way) at any time. More often we are interested in the maximum amplitude reached by the waveform. For the domestic a.c. mains supply, the maximum value of the a.c. voltage is about 340 V, not 240 V which is its root mean square (r.m.s.) value. The r.m.s. value is the equivalent d.c. voltage which would provide the same power as the a.c. supply. Thus the r.m.s. value is a sort of average value of the varying voltage. The period and frequency of an a.c. waveform are related to each other. The **period** is the time between two consecutive maximum values. For the mains a.c. supply, the period T is 1/50 of a second (0.02 s).

The **frequency** is the number of complete periods of the waveform in one second. For the mains a.c. supply, the frequency f is equal to 50 hertz (50 Hz). The relationship between frequency and period is the simple equation $f = 1/T$. So what is the frequency of the a.c. waveform shown in Fig. 4.2b?

Electrical signals that have sinusoidal and square waveforms are common in electronics. The signals delivered by a microphone in a recording studio comprise a mixture of sinusoidal waveforms. These are called audio frequency signals since they are detectable by the human ear. Their frequencies fall in the range 20 Hz to 20 kHz, but mature people are unlikely to hear frequencies above about 10 kHz. Sounds having frequencies above 20 kHz are known as ultrasonic sounds, and these are heard by dogs, bats and other animals. Sounds having frequencies below about 10 Hz are inaudible and known as infrasound. A person can be killed by the vibrations induced in the body by high-intensity infrasound.

4.2 Large and small numbers

In electronics, values of currents and potential differences are generally small and values of resistances are large. For example, the value of a resistance might be one million ohms (1 000 000 Ω). Instead of writing down all the zeros for this large number, it is much easier to use the prefix 'M' meaning 'mega' for one million and write the resistance as 1 MΩ. Similarly, small values of current can be expressed using the prefix 'm' meaning 'milli' for 'one thousandth of', or 'μ' meaning 'micro' for 'one millionth of'. Thus ten milliamperes can be written as 10 mA, and 100 microamperes as 100 μA. The table opposite summarises the values of some of the prefixes you can expect to meet in dealing with electronic circuits.

There is also a shorthand way of doing calculations which involve large and small values. This is done by expressing the values as 'powers of ten' rather than by writing down a lot of zeros. The powers of ten which are equivalent to the prefixes are also listed in the table. Thus the factor 1000 meaning 'one thousand times' is expressed as 10^3, meaning 'ten to the power three', i.e. $10 \times 10 \times 10 = 1000$. And the factor 0.000001 meaning 'one millionth of' is expressed as 10^{-6}, meaning 'ten to the power minus 6', i.e. $1/(10 \times 10 \times 10 \times 10 \times 10 \times 10)$. The numbers above the tens are called indices. When large and small numbers are expressed as powers of ten, calculations become easier because the indices can be added together.

Prefix	Factor	Powers of ten	Symbol
tera	1 000 000 000 000	10^{12}	T
giga	1 000 000 000	10^{9}	G
mega	1 000 000	10^{6}	M
kilo	1 000	10^{3}	k
milli	0.001	10^{-3}	m
micro	0.000 001	10^{-6}	μ
nano	0.000 000 001	10^{-9}	n
pico	0.000 000 000 001	10^{-12}	p

The values of some prefixes

Let's use numbers involving powers of ten in a sample calculation based on the equation $V = I \times R$ (Section 3.3). Suppose you want to work out the value of potential difference across a component which has a resistance of four thousand seven hundred ohms (i.e. 4700 ohms, or 4.7 kΩ) when a current of two milliamperes (i.e. 0.002 A, or 2 mA) flows through it. Now 4.7 kΩ can be written as 4.7×10^{3} Ω, i.e. 'four point seven times ten to the power three', where $10^{3} = 10 \times 10 \times 10$. And 2 mA can written as 2×10^{-3}, i.e. 'two to the power minus 3', where $10^{-3} = 1/(10 \times 10 \times 10)$. Thus the equation gives

$$V = I \times R = 2\,\text{mA} \times 4.7\,\text{k}\Omega$$
$$= 2 \times 10^{-3} \times 4.7 \times 10^{3}\ \text{volts}$$
$$= 9.4\,\text{V}.$$

The calculation is simplified since the indices can be added (3 added to −3 equals zero). And $10^{0} = 1$.

Now take a second example. Suppose you need to know the current flowing through a component which has a resistance of 3.3 MΩ (i.e. 3 300 000 Ω) when there is a potential difference of 4 V across it. We need the equation $I = V/R$ (Section 3.3). The resistance can be written as 3.3×10^{6} Ω, so that $I = 4\,\text{V}/3.3 \times 10^{6}$ Ω. Now $1/10^{6} = 10^{-6}$, since taking ten to the power of something from below the line to above the line changes the sign of the 'power'. Thus $I = 1.2 \times 10^{-6}$ amperes = 1.2 μA.

4.3 The multimeter

Circuit designers could not test and develop new components and circuits without the use of multimeters and oscilloscopes. These instruments measure current, potential difference and resistance, and are able to examine the shapes of waveforms. Broadly speaking, a multimeter is designed to measure d.c. electrical quantities, and an oscilloscope a.c. electrical quantities. Thus a multimeter of the type shown in Fig. 4.3 combines in one portable, battery-operated instrument the facilities for measuring current, potential difference and resistance. These measurements are made on a number of selectable ranges covering both small and large values, e.g. ranges having maximum values of 10 mA and 1000 V. When this multimeter is used as an ammeter to measure current, it is placed in series with a component as shown in Fig. 4.4a. The resistance of the ammeter must be small if it is to have a negligible effect on the value of the current being measured. But when the multimeter is used as a voltmeter to measure volts, it is connected in parallel with a component as shown in Fig. 4.4b. Since a voltmeter should not disturb the p.d. being measured, it must have a very high resistance.

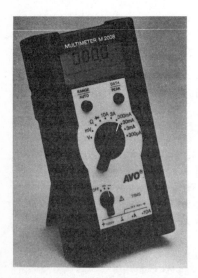 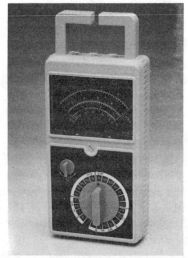

Fig. 4.3   Two types of multimeter: left, digital; right, analogue
Courtesy: Megger Instruments Ltd

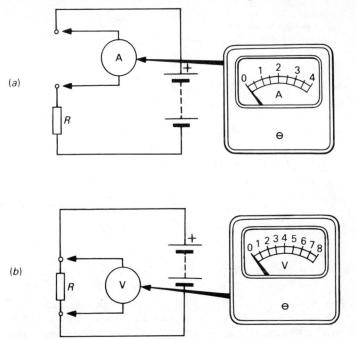

Fig. 4.4 Using a multimeter (a) as an ammeter, and (b) as a voltmeter

This change of the resistance when switching the multimeter from 'volts' to 'amps' is achieved very simply. For the measurement of current, low value resistors are automatically connected in parallel with meter terminals; this ensures that most of the current flows through this low resistance path leaving a calculated small amount to operate the meter movement. For the measurement of voltage, high value resistors are automatically connected in series with one of the meter terminals; this ensures that most of the applied voltage is 'dropped' (see Chapter 5) across these resistors leaving a calculated small amount to operate the meter movement.

When a multimeter is being used as an ohmmeter to measure the resistance of a component, it brings into action an internal battery which makes a small current flow through the component. As shown in Fig. 4.5, an ohmmeter produces a small current which flows through the component under test, and the display records this current. The smaller the current, the larger the resistance of the component. This is why the resistance scale on an analogue multi-

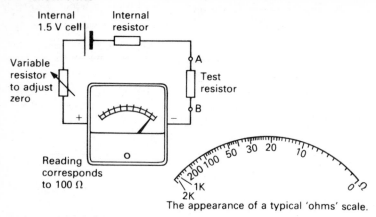

The appearance of a typical 'ohms' scale.

Fig. 4.5 An analogue ohmmeter: (a) the internal circuit, and (b) a typical 'ohms' scale

meter has the zero of the resistance scale on the right of the display. What do you think is the resistance reading at the left of the scale? A digital multimeter simply displays a number giving the value of the resistance. Note that a component's resistance should not be measured when it is in a circuit, since it has other components connected to it and they may affect the resistance being measured.

An analogue multimeter has moving parts comprising a coil of wire which twists in a strong magnetic field. The angle of twist is proportional to the strength of the current flowing through the coil. Attached to the coil is a pointer which moves over a graduated scale. A digital multimeter does not have any moving parts, and what it measures is displayed on a liquid crystal display (LCD), or a light-emitting diode (LED) display. Digital multimeters, in common with many other digital instruments, are less likely to be damaged by mechanical shocks and incorporate 'user-friendly' facilities such as automatic ranging and a clear indication of the type of signal being measured. By plugging in sensors, a digital multimeter can readily be used to measure quantities such as frequency and temperature.

A digital multimeter performs rather better than an analogue multimeter when called upon to measure voltage. This is because a digital voltmeter draws less current from the circuit under test and therefore has less effect on the voltage being measured. This effect is given by the multimeter's sensitivity. Thus a general-purpose analogue multimeter might have a sensitivity of 20 kΩ per volt (20 kΩ V^{-1}), or 100 kΩ V^{-1}, while a general-purpose digital multimeter could have a sensitivity of at least 1 MΩ V^{-1}. The sensitivity is inversely related to the current drawn from the circuit under test. Thus when a meter with a sensitivity 20 kΩ V^{-1} measures a p.d. of 1 V, it draws a current of 1 V/20 kΩ or 50 μA from the circuit. This compares with 1 V/1 MΩ or 1 μA for the digital multimeter. But remember that for a particular sensitivity, the higher the voltage range selected the greater the resistance of the meter and the less it 'loads' the circuit under test. Thus the 20 kΩ V^{-1} voltmeter has a resistance of 20 kΩ on the 1 V range and a resistance of 200 kΩ on the 10 V range.

4.4 The oscilloscope

The oscilloscope is a versatile instrument and it is used to measure the characteristics of both d.c. voltages and a.c. voltages. An oscilloscope like the one shown in Fig. 4.6 can display on its screen part of a rapidly changing waveform so that measurements can be made of the frequency, shape and period of the waveform. It is therefore widely used in the design and development of amplifiers, music synthesisers, televisions, radios and computers. Oscilloscopes are generally operated from the mains power supply for use in laboratories and workshops. Portable battery-operated oscilloscopes are available for engineers and scientists working outdoors.

The full name for an oscilloscope is cathode-ray oscilloscope (CRO) since its main component is a **cathode-ray tube** (CRT). The

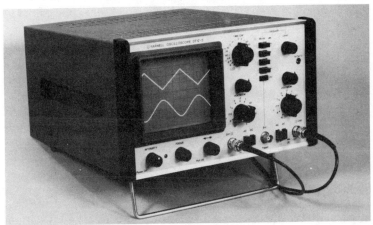

Fig. 4.6 An oscilloscope
Courtesy: Farnell Instruments Ltd

CRT is also used in televisions, visual display units and radar sets. It incorporates an electron gun which 'fires' a single narrow beam of rapidly moving electrons (cathode rays) at a phosphor coated screen which glows at the point where the beam strikes it. By moving the electron beam in sympathy with the signal waveform, it is possible to 'write' a graph (a trace) of the waveform on the screen. A dual-trace oscilloscope enables the shape of two waveforms to be written on the screen simultaneously. This is achieved by electronic circuits that rapidly move the electron beam to separate parts of the screen to give the impression that there are two independent beams. This enables the comparison of the shapes of two waveforms. The more costly dual-beam oscilloscopes have two electron guns and they can produce the shape of two higher frequency waveforms simultaneously.

The inside surface of the screen of a cathode-ray tube is coated with a phosphor which lights up, or fluoresces, when electrons strike it. Different phosphors produce different colours of light, though green is favoured for most 'scopes. Of course, once the phosphor has been activated by the impact of the electrons, its light must fade away within a few milliseconds if the trace is not to linger and confuse the picture. However, some cathode-ray tubes are designed to capture the shape of a single waveform so that it can be examined easily. For example, an electrocardiograph uses a cathode-ray tube which retains the waveform of heart beats for a few seconds; and so does a radar screen which marks the position of aircraft by the 'echo' of radio waves which bounce off them. Note that the cathode-ray tube in a colour television uses red, green and blue phosphors to create the various colours required to reproduce a colour picture.

Fig. 4.7 shows the inner workings of the type of cathode-ray tube used in oscilloscopes. It has three main parts: an electron gun, a deflection system and a fluorescent screen, all housed in an evacuated glass envelope. This type of tube uses high voltages on the deflecting plates to change the path of the electron beam and to focus the beam into a small spot on the screen. This technique is called electrostatic deflection, as opposed to magnetic deflection which uses the magnetic field produced by a current passing through coils wound round the tube, used, for example, in most television picture tubes. (See Chapter 16).

The electron gun comprises a heated tungsten filament within a nickel cathode cylinder coated with oxides of barium and strontium which give off electrons – a process known as **thermionic emission**, which was widely used in the (now almost obsolete) valve. The

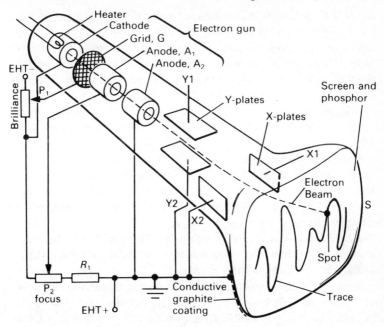

Fig. 4.7 The operation of a cathode-ray tube

electrons are negatively charged so they accelerate towards the anodes A_1 and A_2 which are at a more positive voltage than the cathode. The strength of the electron beam, and hence the brightness of the image on the screen S, is controlled by the potentiometer P_1, which makes the grid more or less negative with respect to the cathode. The rest of the electron gun consists of accelerating and focusing anodes which are shaped metal cylinders, all held at a positive voltage which can be varied to alter the size of the spot produced on the screen. For a small CRT, the positive voltage on the focusing anodes is between 500 V and 1000 V. It is therefore dangerous to fiddle about with the circuits inside an oscilloscope – or a television, for that matter – unless you know what you are doing! A graphite coating inside the CRT avoids the build-up of electrical charge on the screen by collecting any secondary electrons given off by the screen.

After leaving the electron gun, the electron beam enters the deflection system which consists of two sets of metal plates, X and Y, at right angles to each other. The voltage applied to the X-plates is generated by a **time-base** circuit. Its job is to deflect the electron beam horizontally to make the spot 'sweep' across the screen from left to right at a steady speed. The speed can be adjusted by the time-base controls on the oscilloscope. After each sweep, the time-base amplifier switches off the beam and sends it back to the starting point at the left end of the screen – this process is known as flyback.

The waveform to be examined on the CRO is amplified and applied to the Y-plates. The amplification of the waveform can be adjusted by the Y-sensitivity controls. The input waveform causes the horizontal trace to move vertically in response to the strength of the waveform. A stable trace appears on the screen when each horizontal sweep of the trace starts at the same point on the left of the screen. This is achieved by feeding part of the input waveform to a trigger circuit. This starts, i.e. triggers, the time-base circuit when the input signal has reached a particular amplitude set by the trigger level control. Most CROs allow manual and automatic adjustment of the triggering of the time-base.

In order to make measurements of the amplitude and frequency of a signal, a scale marked out as a grid of lines spaced at 10 mm (1 cm) intervals covers the screen of a CRO. Thus, when a rectangular a.c. waveform is fed to the oscilloscope, the trace on the screen might appear as shown in Fig. 4.8. Suppose this trace occurs when the Y-sensitivity control is set at 2 volts per cm (2 V cm^{-1}). Clearly the amplitude of the signal is about 3 V. The frequency of the signal is obtained from the time-base setting. Suppose this setting is 1 millisecond per 10 mm (1 ms cm^{-1}). Now the period of the signal is the distance marked T and this occupies a distance of 5

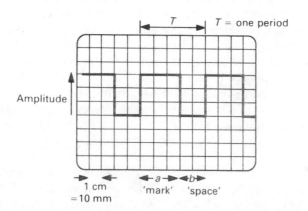

Fig. 4.8 A typical trace on the screen of an oscilloscope

cm on the scale. This represents a time of 1 ms × 5 = 5 ms. And since the frequency f of the waveform is equal to $1/T$, the signal has a frequency of 1/5 ms = 200 Hz. The scale also gives the mark-to-space (M/S) ratio of the signal. This is the ratio of the times the signal is HIGH to when it is LOW, i.e. the ratio $a:b$. From the scale the M/S ratio is 3:2.

4.5 Black boxes

If you are new to electronics, it is not a good idea to begin by studying the way individual devices work. If you do, you could easily get bogged down in the detail of how the devices work and lose sight of the interesting ways in which they are used, e.g. in measurement, communications, control and computing as explained in Chapter 1. We want to keep these real world applications in the forefront of our minds. To make sure we do, it is better to focus on what electronic devices do rather than on how they work.

For example, an amplifier is a common enough device in electronics. Amplifiers are used in radios, televisions and communications satellites, indeed anywhere it is necessary to increase the strength of a signal. In order to understand the overall purpose of an amplifier, we simply need to know what the amplifier does, not how it works. So in diagrams the amplifier is drawn as a 'black box' as shown in Fig. 4.9. It is called a black box not because of its colour (you can colour it any way you like!) but because we are interested in what it does, not what goes on inside it. ('Black' because its contents are inscrutable, perhaps?) The black box has information going into it (called the **input**) which it acts upon in some way to produce information leaving it (called the **output**). In electronics a black box is an *activity* box, since what comes out of it is different from what goes into it. Thus the output of an amplifier is information that is in some way stronger than the input. Sensors, counters, memories and displays are some of the many black boxes used in electronics. When black boxes such as these are connected together in a purposeful way, an electronic system is created.

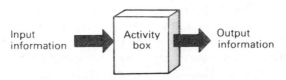

Fig. 4.9 A black box amplifier

4.6 Electronic systems

The thermometer shown in Fig. 4.10 is an example of an electronic system for measuring temperature. This instrument may be regarded as made up of three black boxes as shown by the systems diagram in Fig. 4.11. The sensor (black box 1) is placed in contact with the thing you want the temperature of. The information produced by the sensor is input to the amplifier (black box 2). The amplifier outputs information suitable for operating the display (black box 3). A systems diagram like this simplifies a complex function, e.g. measuring temperature, and it helps to show what part each black box plays in the operation of the system even though you may not be sure how the black boxes achieve what they do. Chapter 15 describes the design of an electronic thermometer based on these three black boxes.

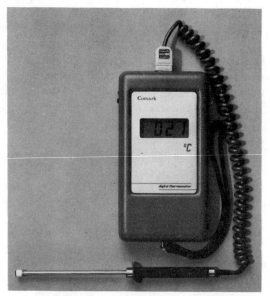

Fig. 4.10   A digital thermometer
Courtesy: Comark Electronics Ltd

A systems diagram for a radio receiver is shown in Fig. 4.12. It comprises a tuned circuit (black box 1) which selects the narrow band of radio frequencies used to carry a message from a transmitting station. These radio frequencies are amplified by a radio frequency amplifier (black box 2). A detector and filter (black box 3) then converts the amplified message into a form suitable for

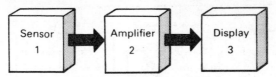

Fig. 4.11 The electronic system of a thermometer

operating an earpiece. Further amplification by the audio amplifier (black box 4) enables a loudspeaker to be operated (black box 5). A practical design for a radio receiver based on these four black boxes is described in Chapter 16.

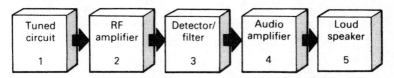

Fig. 4.12 The electronic system of a radio receiver

Fig. 4.13 shows the systems diagram for a microcomputer. A microprocessor (black box 1) carries out a list of instructions held in a memory (black box 2). The input ports and the output ports (black box 3) are the microcomputer's 'windows' to the outside world through which information comes and goes. Within the microcomputer, information travels along electrical conductors called the 'highway' (or bus) which connects together the various parts of the microcomputer system.

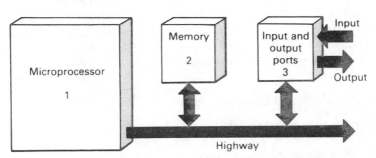

Fig. 4.13 The electronic system of a microcomputer

4.7 Systems and subsystems

The systems diagrams above can show as much or as little detail as we like. For example, Fig. 4.14 shows the systems diagram for a

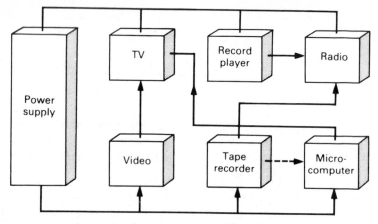

Fig. 4.14   An entertainment system

home entertainment system. A number of functions could be interconnected (dotted line) to produce the particular entertainment we want – though we are unlikely to use them all at once! For example, we could use the cassette recorder to store programs from the microcomputer. Or the television to play back recordings made on the portable video recorder.

Each individual function in the entertainment system is known as a **subsystem**. It is not necessary to show the detail of the working of each subsystem, for that would confuse our general plan for the entertainment system. But if we were inclined to show this detail, we could include a systems diagram for one or more of the subsystems, e.g. the radio receiver (Fig. 4.12). Further detail might show the amplifier as a set of subsystems, i.e. its preamplifier, tone control, etc. At a still finer level of detail, individual components could be shown as the subsystems of the tone control.

4.8   Analogue and digital systems

The words 'analogue' and 'digital' are in general use for describing electronic devices. For example, electronic watches are said to be analogue or digital according to the type of display they use. The word 'analogue' means 'model of', so an analogue watch models the smooth passage of time by using hands which move smoothly round its face. The advantage of this type of display is that it is easy to get an idea of the present time in relation to past and future time as displayed on the face. The word 'digital' means 'by numbers', so the

Fig. 4.15 An analogue photographic exposure meter
Courtesy: Weston Instruments

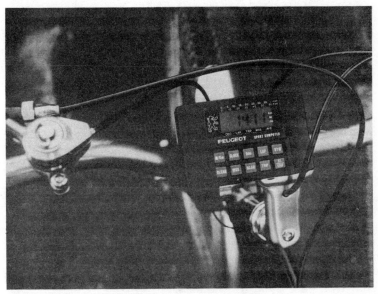

Fig. 4.16 A 'Sports Computer'
Courtesy: Cycles Peugeot

digital watch displays the current time as a set of numbers which change abruptly at intervals.

Some electronic devices combine both analogue and digital functions. Thus an analogue watch uses digital circuits for timing and then displays the time in analogue form. It is interesting to note that spring-driven watches work digitally, in that the escape mechanism monitors time as a series of 'ticks'. The photographic exposure meter shown in Fig. 4.15 is an analogue system from start to finish. Its sensor produces an electrical signal which is an analogue of the smooth variation in light intensity from dawn to dusk, or from indoors to outdoors. This electrical analogue of the changing light intensity is passed on to an amplifier before operating a pointer moving over a scale.

The 'Sports Computer' shown in Fig. 4.16 is an entirely digital system. This instrument is used for recording a number of useful events when cycling, such as speed, average speed, maximum speed, distance and, of course, time. A sensor on a wheel monitors each rotation of the wheel and produces a series of on/off pulses. These pulses are fed to counters and other digital circuits inside the device to display information in digital form.

5

Potential Dividers and Resistors

5.1 The basic potential divider

Fig. 5.1a shows the function of this useful black box. It provides an output potential difference, V_{out}, which is less than the input potential difference, V_{in}. But why should this be a useful function? Well, potential dividers are used as volume controls in radios, and for controlling the brightness of television screens. And, as you will see, potential dividers are widely used in other circuit designs, e.g. in control and instrumentation systems where a p.d. has to be reduced to a value suitable for operating transistors and integrated circuits.

Fig. 5.1b shows a circuit that acts as a potential divider by reducing the e.m.f. of a battery. The two rectangular symbols, marked R_1 and R_2, are electronic components called **resistors**; they simply have an electrical resistance measured in ohms. Various types of resistors are described in Section 5.3. The reduced output

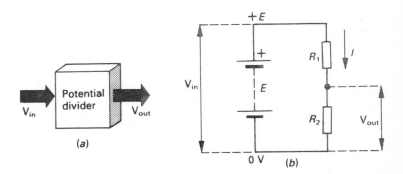

Fig. 5.1 (a) The function of a potential divider, and (b) the use of two resistors as a potential divider

p.d., measured with respect to 0 V, occurs at the join between the two resistors. Any value of p.d. between 0 V and V_{in}, the e.m.f. of the battery, can be obtained by changing the values of the resistors. The potential divider works equally well for reducing a p.d. in alternating current circuits.

The values of the resistors R_1 and R_2 determine the output p.d. V_{out}. The equation is

$$V_{out} = V_{in} \times \frac{R_2}{R_1 + R_2}$$

This equation shows that V_{out} is less than V_{in} by the fraction $R_2/(R_1 + R_2)$, i.e. the smaller R_2, the smaller V_{out}. Just suppose that $V_{in} = 9$ V, and the values of the resistors are $R_1 = 90 \ \Omega$ and $R_2 = 10 \ \Omega$. Now

$$V_{out} = 9 \text{ V} \times \frac{10}{100} = 0.9 \text{ V}.$$

The same value for V_{out} could have been obtained if $R_1 = 900 \ \Omega$ and $R_2 = 100 \ \Omega$, for then

$$V_{out} = 9 \text{ V} \times \frac{100}{1000} = 0.9 \text{ V}.$$

Or if $R_1 = 240 \ \Omega$ and $R_2 = 120 \ \Omega$, then

$$V_{out} = 9 \text{ V} \times \frac{120}{360} = 3 \text{ V}.$$

Thus, it is the ratio of the values of R_1 and R_2 which determines the output p.d. of a potential divider, not their actual values. The above equation can be proved very simply by using Ohm's law (Section 3.3.). First note that the current I flows through both resistors, and is equal to

$$I = \frac{V_{in}}{R_1 + R_2}$$

And V_{out} is given by $V_{out} = I \times R_2$.

If we substitute I from the first equation into the second,

$$V_{out} = V_{in} \times \frac{R_2}{R_1 + R_2}$$

as required.

5.2 Resistors in series and parallel

The combined resistance of two resistors connected in series is found by adding their values. Thus in Fig. 5.2a the total resistance R of two resistors R_1 and R_2 connected in series is given by the equation

$$R = R_1 + R_2$$

This can easily be proved. First note that when two resistors are connected in series the same current, I, flows through each resistor. Second, the sum of the p.d.s across the two resistors is equal to the p.d. across the combination. Thus $V = V_1 + V_2$. And since $V = I \times R$, we can write

$$V = IR_1 + IR_2 = I(R_1 + R_2) = IR$$

Here we have written $R = R_1 + R_2$ for the resistance of the combination. Thus if we replace the two resistors connected in series by a single resistor equal to the sum of the combination, the current drawn from the battery remains unaltered.

Now when two resistors are connected in parallel as shown in Fig. 5.2b, the total resistance of the combination is given by the equation

$$\frac{1}{R} = \frac{1}{R_1} + \frac{1}{R_2}$$

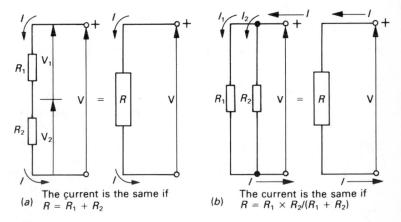

(a) The current is the same if $R = R_1 + R_2$

(b) The current is the same if $R = R_1 \times R_2/(R_1 + R_2)$

Fig. 5.2 (a) Two resistors connected in series, and (b) two resistors connected in parallel

which can be rewritten

$$R = \frac{R_1 \times R_2}{R_1 + R_2}$$

This equation can be proved by first noting that the p.d. across each resistor is equal to the p.d. across the combination. Second, the sum of the currents through each resistor is equal to the current flowing from the power supply. Thus $I = I_1 + I_2$. And since $I = V/R$, and $I_1 = V/R_1$ and $I_2 = V/R_2$, we can write

$$I = \frac{V}{R} = \frac{V}{R_1} + \frac{V}{R_2}$$

which reduces to

$$\frac{1}{R} = \frac{1}{R_2} + \frac{1}{R_2}$$

Note that when two or more resistors are connected in series, their total resistance is *more* than the largest value present. But when two or more resistors are connected in parallel, their total resistance is *less* than the smallest value present. We can prove this by taking two values for R_1 and R_2, e.g. let $R_1 = 300\,\Omega$ and $R_2 = 500\,\Omega$. If these two resistors are connected in series, their combined resistance is $300 + 500 = 800\,\Omega$, which is more than the largest value present, i.e. more than $500\,\Omega$. And if they are connected in parallel, their combined resistance is

$$\frac{300 \times 500}{300 + 500} = \frac{150\,000}{800} = 187.5\,\Omega$$

which is less than the smallest value present, i.e. less than $300\,\Omega$.

5.3 Fixed-value and variable resistors

Fig. 5.3 shows examples of fixed-value resistors and their circuit symbol. The carbon-film resistor is made by depositing a hard crystalline carbon film on the outside of a ceramic rod, and then protecting it by a hard-wearing, electrically-insulating coating. The resistance of the carbon film between the connecting wires is the resistor's value. Of similar construction is the metal-film resistor, except that tin oxide replaces the carbon. The metal-film resistor has better temperature stability than the carbon-film resistor. Both types are recommended for use in audio amplifiers and radio receivers where they are exposed to extreme changes of temperature and humidity. These resistors also generate little electrical

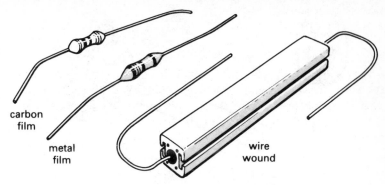

Fig. 5.3 Examples of fixed-value resistors

'noise'. (Electrical noise is the 'hiss' which tends to 'drown' the required signal in a circuit, and is caused by the random movement of electrons in the resistor.) Wire-wound resistors are made by winding a fine wire of nichrome (an alloy of nickel and chromium) round a ceramic rod. A wire-wound resistor can be made to have a very precise value which can be guaranteed to within 0.1%.

The thick-film resistor is made by adjusting the thickness of a layer of semiconducting material to give the required resistance. These resistors are generally grouped eight at a time in a single-in-line or dual-in-line package (Fig. 5.4). The individual resistors may be independent of each other or have a common connection, depending on the application. Thick-film resistors of this type are particularly useful in computer circuits where eight or more connections have to be made between the computer and display or control circuits.

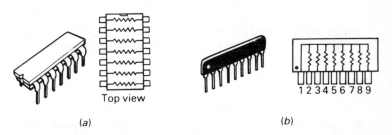

Fig. 5.4 Resistors grouped together as (a) a dual-in-line (d.i.l.) package, and (b) a single-in-line (s.i.l.) package

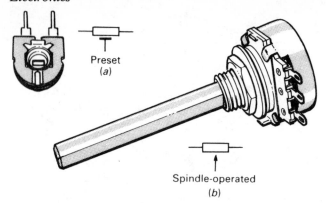

Circuit symbols

Fig. 5.5   Types of variable resistor: (a) preset type, and (b) spindle-operated

Fig. 5.5 shows the two main types of variable resistor, together with their circuit symbols. They each have three terminals: two make contact with the ends of a carbon or wire-wound resistive track and the third is attached to a 'wiper' which moves over the track. The resistance between one end of the track and the wiper varies as the wiper moves along the track. The three terminals enable the variable resistor to act as a potential divider (Section 5.1) since effectively it comprises two variable resistors connected in series. As the resistance of one resistor increases, the other decreases. When a variable resistor is used as a potential divider, it is known as a **potentiometer**. Spindle potentiometers (or 'pots') are intended to be mounted on a panel and they are then adjusted by a knob fixed to the spindle. The 'preset' type is generally soldered in place in a circuit and then adjusted just once before being left alone. These preset variable resistors are set using a small screwdriver or special adjusting tool. Some preset variable resistors can be adjusted with great precision, and are classed as 10-turn or 20-turn presets depending on how many turns of the adjusting screw are required to move the wiper from one end of the resistance track to the other.

5.4 Values and codings of resistors

Most fixed-value resistors are marked with coloured bands so that their values can be read easily. Four coloured bands are used as shown in Fig. 5.6.

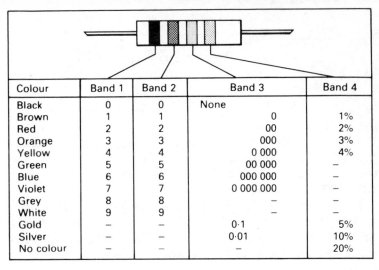

Colour	Band 1	Band 2	Band 3	Band 4
Black	0	0	None	
Brown	1	1	0	1%
Red	2	2	00	2%
Orange	3	3	000	3%
Yellow	4	4	0 000	4%
Green	5	5	00 000	–
Blue	6	6	000 000	–
Violet	7	7	0 000 000	–
Grey	8	8	–	–
White	9	9	–	–
Gold	–	–	0·1	5%
Silver	–	–	0·01	10%
No colour	–	–	–	20%

Fig. 5.6 The resistor colour code

Band 1 gives the first digit of the value.

Band 2 gives the second digit of the value.

Band 3 gives the number of zeros which follows the first two digits.

Band 4 gives the 'tolerance' of the value worked out from the first three bands.

For example, suppose the bands are coloured as follows:

Band	1	2	3	4
Colour	yellow	violet	red	silver
Value	4	7	00	10%

This resistor has a value of 4700 Ω to within 10% more or less. That is, its value is 4.7 kΩ ± 10%. If the value of this resistor were measured accurately, its resistance should not be more than 4.7 kΩ + 0.47 kΩ, or less than 4.7 kΩ − 0.47 kΩ, i.e. between 5.17 kΩ and 4.23 kΩ. For most circuit designs, it is unnecessary to use resistors with tolerance better than 5%.

Instead of using a colour code, some manufacturers are marking the values of resistors using the British Standards BS1852 code. This code is often used to mark resistor values on circuit diagrams, as in this book. The BS1852 code consists of letters and numbers as the following examples show.

BS1852 code	Resistance
6K8M	$6.8\,k\Omega \pm 20\%$
R47K	$0.47\,\Omega \pm 10\%$
5R6J	$5.6\,\Omega \pm 5\%$
47KG	$47\,k\Omega \pm 2\%$
2M2F	$2.2\,M\Omega \pm 1\%$

Note that in the BS1852 code, instead of the decimal point a letter, e.g. 'K', is used to indicate the multiplying factor. Thus in the code 6K8M, the K indicates that the resistor has a value of $6.8 \times 1000\,\Omega = 6.8\,k\Omega$.

To avoid making an impossibly large number of resistor values, manufacturers produce only certain values known as preferred values. For 10% tolerance resistors, the preferred values belong to the so-called E12 series in which the twelve values are:

$$10, 12, 15, 18, 22, 27, 33, 39, 47, 56, 68, 82.$$

And for 5% tolerance resistors, the E24 series includes an additional twelve values as follows:

$$11, 13, 16, 20, 24, 30, 36, 43, 51, 62, 75, 91.$$

5.5 Power ratings of resistors

Heat is produced (or dissipated) within a resistor when current flows through it. This heat is generally a nuisance in electronic circuits, but it is put to good use in electric soldering irons, electric lamps and fire bars, and the like. The heat represents the conversion of electrical energy into heat energy. Now like all forms of energy, heat energy is measured in joules. The rate at which heat is produced within the resistor is measured in joules per second which is equal to the power produced in the resistor measured in watts. Thus

power (watts) = rate energy is produced (joules per second)

or $\qquad W = J\,s^{-1}$

The unit of electrical power, the watt, is used to estimate the power rating of all types of devices, from soldering irons (e.g. 20 W) to power station generators (e.g. 10 MW), from solar panels on spacecraft (e.g. 2 kW) to resistors (e.g. 250 mW). For a resistor, it is a simple matter to calculate the electrical power dissipated within it as heat. It is given by the equation

$$W = V \times I$$

where V is the p.d. across the resistor and I is the current flowing through the resistor.

So suppose a current of 10 mA flows through a resistor when the p.d. across it is 5 V. The heat generated within the resistor is then

$$5 \text{ V} \times (10/1000) \text{ A} = 0.05 \text{ W or } 50 \text{ mW}$$

Note that it is usual to state the electrical power generated within resistors, and other components, in milliwatts (mW) since it is usually less than 1 W. Now to save you the trouble of measuring the current flowing through the resistor, it is better to use an alternative form of the power equation. You may remember that $V = I \times R$ (Section 3.3). So $I = V/R$ and the power equation becomes

$$W = V \times I = V \times V/R = V^2/R$$

These two quantities, V and R, are readily obtained; R from the colour code on the resistor, or from an ohmmeter, and V from a voltmeter. If you do happen to know the value of the current flowing through the resistor, then use the following alternative form of the power equation

$$W = V \times I = I \times R \times I = I^2 \times R$$

Resistors, both fixed-value and variable, are rated according to the maximum allowable power generated in them. Exceed this rating, and the resistor is likely to be damaged by its own self-heating. And the same precaution applies to many other electronic devices, such as transistors and diodes, which have a maximum safe power rating. General-purpose resistors are rated at 1/4 W (250 mW), 1/2 W (500 mW), 1 W, and 2 W. For most of the applications discussed in this book, 250 mW carbon-film or metal-film resistors are suitable, though there is no harm in using resistors of higher rating.

Suppose a 1 kΩ resistor is used in circuit where the p.d. across it is 16 V. The power dissipated in the resistor is

$$W = V^2/R = (16 \text{ V})^2/1\text{k}\Omega = 256/10^3 = 256 \text{ mW}$$

Thus a 250 mW resistor is just adequate and you would expect it to be warm to the touch. A 500 mW type would have more than adequate capacity to dissipate this amount of heat.

5.6 Special types of resistor

Some resistors change their resistance in response to the change in some property. Two of the most useful devices of this type are shown in Fig. 5.7. The resistance of the **light-dependent resistor** (LDR) changes with the amount of light falling on it. The resistance of the **thermistor** changes with temperature. The graphs show how an ohmmeter would record the change in resistance of these devices. The resistance of the LDR would increase greatly from daylight to darkness, while the resistance of the thermistor would increase by a much smaller amount if its temperature fell from, say, 100 °C to 0 °C. Both the LDR and the thermistor are based on semiconductors. The LDR uses the material cadmium sulphide, and the thermistor a mixture of different semiconductors. While most LDRs are like the one shown in Fig. 5.7a, thermistors may be disk-shaped (as shown in Fig. 5.7b) or rod-shaped. For maximum sensitivity, a small piece of semiconductor may be encapsulated in a glass envelope. This 'glass bead thermistor' is ideal for temperature measurement since it responds quickly to temperature changes.

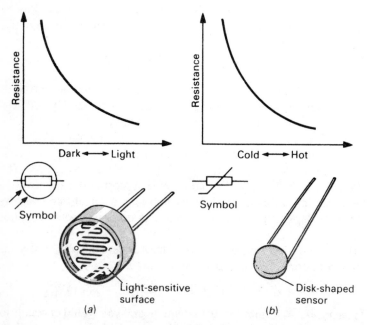

Fig. 5.7 Two special types of resistor: (a) the light-dependent resistor, and (b) the thermistor

Light-dependent resistors are used in photographic lightmeters, security alarms and automatic street light controllers. Thermistors are used in thermostats, fire alarms and thermometers.

The LDR and thermistor are generally used as one of the resistors in a potential divider as shown in Fig. 5.8, so that a change of resistance produces a change of p.d. Thus, if we use the following relationship (Section 5.1)

$$V_{out} = V_{in} \times \frac{R_2}{R_1 + R_2}$$

where R_2 is the resistance of the LDR or thermistor, it is easy to see what happens when the resistance of the LDR or thermistor changes. Fig. 5.8a shows that when the LDR is in shade, its resistance, R_2, is high so that the p.d. across it (i.e. V_{out}) is high. If it is in sunlight its resistance falls, so V_{out} falls.

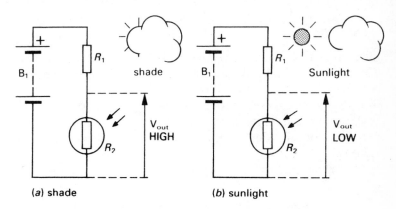

(a) shade   (b) sunlight

Fig. 5.8 Using the LDR in a potential divider

Similarly, if the thermistor replaces the LDR in this potential divider, high temperature makes the resistance, R_2, of the thermistor low and so V_{out} is low. A low temperature increases the thermistor's resistance so that V_{out} is high. Note that there is a smooth change of output voltage from the potential divider with change of light intensity or temperature. Thus the potential divider is an analogue device, since V_{out} changes smoothly with change of temperature or light intensity. The potential divider in which one resistor is an LDR or a thermistor is a very useful arrangement in circuits – see Chapters 8 and 14.

5.7 The Wheatstone bridge

The particular combination of a pair of potential dividers shown in Fig. 5.9 is named after its 19th century inventor Sir Charles Wheatstone. (He is also credited with the invention of the electric telegraph and the concertina.) The two potential dividers are connected in parallel across the same power supply, E. Thus the voltage at X, i.e. the p.d. across R_2, is given by

$$\frac{E \times R_2}{R_1 + R_2} = \frac{E}{1 + R_1/R_2}$$

And the voltage at Y, i.e. the p.d. across R_3, is given by

$$\frac{E \times R_4}{R_3 + R_4} = \frac{E}{1 + R_3/R_4}$$

Thus if

$$\frac{R_1}{R_2} = \frac{R_3}{R_4},$$

the voltage at X equals the voltage at Y. There is then no potential difference between the points X and Y, and a voltmeter connected between these two points would not show a deflection. The Wheatstone bridge is said to be 'balanced'. Wheatstone used the bridge to make accurate measurements of resistance, since if three of the resistor values in the above relationship are known, the fourth value can be found.

The Wheatstone bridge is very useful in electronic measurement

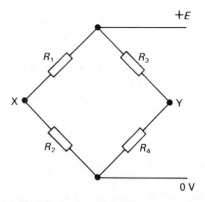

Fig. 5.9   The Wheatstone bridge network of resistors

and control circuits since a voltmeter placed between the points X and Y is very sensitive to any changes in the resistor values. Suppose the voltmeter is a centre-zero type as shown in Fig. 5.10, so that it responds to any difference of voltage between X and Y. In practice, one of the resistors, R_1 say, might have a resistance which varies with temperature, light intensity or pressure. In Fig. 5.10, a temperature-sensitive thermistor, Th_1, takes the place of R_1. At a particular temperature, the Wheatstone bridge is first balanced by adjusting the value of VR_1 so that the voltmeter reads 0 V. Now if the temperature rises, there is a decrease in the resistance of the thermistor and the voltmeter will show a reading. The increase of temperature makes the voltage at point X rise above that at point Y. The voltmeter records this increase of voltage. A cooling of the thermistor causes a decrease in the reading. The voltmeter could, of course, be used as a simple electronic thermometer if its scale were calibrated in degrees Celsius.

The above equation could be used to work out the resistance of the thermistor if the values of VR_1, R_3 and R_4 were known. Thus the above equation can be written

$$R_1 \text{ (resistance of thermistor)} = \frac{R_3 \times R_2}{R_4}$$

where R_2 is the value of VR_1. In order to detect when the bridge is balanced, the voltmeter should be a sensitive millivoltmeter – the 0.05 V (50 mV) range on a multimeter would be suitable though a multimeter would not have a centre-zero reading. The Wheatstone bridge is used for the control systems discussed in Chapter 15.

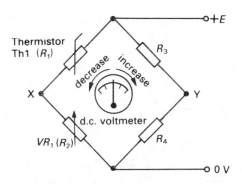

Fig. 5.10 Using a thermistor in a Wheatstone bridge for measuring temperature

5.8 The strain gauge

Strictly speaking, there is no such phenomenon as an irresistible force meeting an immovable object. A force, however small, always causes an object to give way slightly; it distorts or 'strains' in response to the force. For so-called 'rigid' objects, the strain is usually very small so a special type of resistive transducer called a strain gauge is used to measure it. A typical strain gauge is shown in Fig. 5.11a and consists of a metal foil which has a resistance between 60 ohms and 2000 ohms.

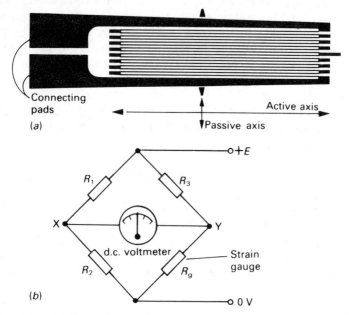

Fig. 5.11   (a) The general appearance of a strain gauge, and (b) how it is used with a Wheatstone bridge

The thin foil is formed by rolling out an electrically resistive material and etching parts away (rather like a printed circuit board is made) leaving a thin flexible resistor in the form of a grid pattern as shown in Fig. 5.11a. The strain gauge is used by gluing it to the surface of the object which is undergoing strain. As the object bends, expands or contracts so does the strain gauge. Now if any metal is stretched, its resistance increases; if it is compressed, its resistance decreases. The resistance change is small, perhaps one tenth of an ohm for a 120 ohm strain gauge. But the Wheatstone bridge can respond to this small change of resistance. Fig. 5.11b shows how the strain gauge is connected in a Wheatstone bridge. One voltage divider comprises resistors R_1 and R_2, and the other comprises resistors R_3 and the strain gauge, R_g. A sensitive voltmeter, V, is placed between X and Y.

As you now know, if $R_1 = R_2$ and $R_3 = R_g$, the voltages at point X and Y are equal and the voltmeter doesn't register a voltage difference. (This balancing of the bridge is usually done by making resistor R_1, R_2, or R_3 a variable resistor.) But if the strain gauge is attached to the surface of a material which stretches (called tensile strain), under the action of a force (called tensile stress), its resistance increases slightly. This makes the voltage at Y rise slightly above the voltage at X and the voltmeter shows a deflection one way. And should the strain gauge be compressed slightly (by a compressive force), it contracts slightly (compressive strain) and its resistance falls. This makes the voltage at Y fall slightly below that at X and the voltmeter shows a deflection in the opposite way. Thus this simple instrumentation circuit shows whether the object to which the strain gauge is attached is bending one way or the other. In practice the small change in voltage between points X and Y is amplified by special instrumentation circuits. These generally use an integrated circuit called an operational amplifier which is described in Chapters 14 and 15.

6

Timers, Oscillators and Capacitors

6.1 What timers and oscillators do

Fig. 6.1 shows the functions of these two black boxes. First the timer: when this device receives a 'trigger' signal at its input, the voltage at its output rises sharply, i.e. becomes HIGH. The output voltage remains HIGH for a time delay of T seconds, and then falls to 0 V again, i.e. the output voltage is then LOW. After the output voltage has fallen to 0 V, the timer needs another trigger signal at its input to repeat the time delay. The time delay is determined by the values of components within the timer black box. Timers are used a lot in electronic systems. Toasters and washing machines, cameras and industrial processes use timers to ensure that an operation, such as exposing a film to light, occurs for a given period of time.

Fig. 6.1b shows the function of one type of oscillator. When this black box is switched on, its output voltage goes HIGH/LOW continually. The waveform of the signals produced by this oscillator

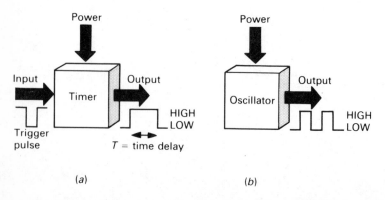

Fig. 6.1   The function of (a) a timer, and (b) an oscillator

is generally known as a rectangular wave. It is a **square wave** if the times for which the output voltage is HIGH and LOW are equal. Oscillators that produce waveforms like this are used in alarm systems for flashing a lamp on and off, or for sounding an audio alarm from a loudspeaker. They are also used in electronic musical instruments, digital clocks and watches and in microcomputers. The function of timers and oscillators is largely determined by the properties of an electronic component called a **capacitor**.

6.2 The way a capacitor works

Fig. 6.2a shows the basic structure of a capacitor, and Fig. 6.2b the circuit symbol which reflects this structure. It comprises two metal electrodes separated by an electrical insulator called a dielectric. The metal electrodes are connected to the terminals of the capacitor. The capacitor is able to store electric charge. If a battery, B_1, is connected across the capacitor as shown in Fig. 6.3a, there is a short flow of electrons in the external circuit from one electrode to the other. Thus one electrode becomes negatively charged and the other positively charged. The p.d. across the terminals is then equal to the e.m.f., E, of the battery and the capacitor is said to be charged. That is, the excess of electrons on one electrode, and the deficiency of an equal number of electrons on the other electrode, represents a store of charge.

If the battery is then removed, these charges remain in place since they are separated by the dielectric – an insulator. Now if the electrodes of the capacitor are joined together by a conductor as

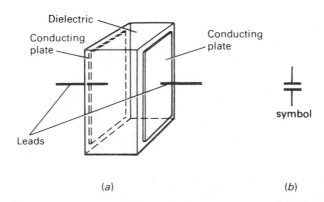

(a)   (b)

Fig. 6.2   (a) The basic structure of a capacitor, and (b) its circuit symbol

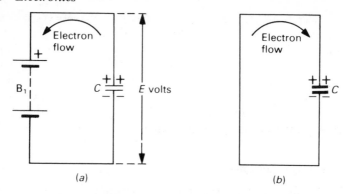

Fig. 6.3   The (a) charging, and (b) discharging of a capacitor

shown in Fig. 6.3b, electrons flow in the reverse direction through the conductor until the p.d. across the capacitor falls to zero. The charge stored by a capacitor for a p.d. of 1 V across its terminals is known as the **capacitance** of the capacitor.

6.3 The unit of capacitance

The unit of capacitance is the **farad** (symbol F). This unit is named in honour of the 19th century English scientist Michael Faraday, to whom we owe so much for the electrified world of today. The farad is defined as follows: it is the capacitance of a capacitor that stores a charge of 1 coulomb when it has a p.d. of 1 V across its terminals. In general, if a charge of Q coulombs is given to a capacitor of C farads and the resulting rise in p.d. is V volts, then

$$Q = C \times V$$

The farad happens to be too large a unit to express the values of capacitors in electronics, and it is necessary to use fractions of a farad as follows (see Section 4.2):

Fraction	Abbreviation
microfarad	μF $(10^{-6}\,\text{F})$
nanofarad	nF $(10^{-9}\,\text{F})$
picofarad	pF $(10^{-12}\,\text{F})$

It is useful to remember that 1000 pF = 1 nF, and 1000 nF = 1 μF. The values of capacitors in common use range from as small as, say, 5 pF to as large as 10 000 μF. Suppose a 100 μF capacitor is used in a circuit where the p.d. across it is 15 V. What is the charge stored by the capacitor? Using the equation $Q = C \times V$ above:

$$Q = 100 \times 10^{-6}\,\text{F} \times 15\,\text{V}$$

$$= 1.5 \times 10^{-3}\,\text{coulombs}$$

$$= 1.5\,\text{millicoulombs}$$

$$= 1.5\,\text{mC}.$$

Note that this is a very small charge but it does represent the movement from one plate to the other of the capacitor of a very large number of electrons – 9.375 thousand million million to be precise! In electronics we are not very concerned with the precise amount of charge a capacitor stores for a given voltage. We are more interested in how the value of a capacitor determines the performance of timers and oscillators. By the way, you should note that the letter 'F' is usually omitted from capacitor values on circuit diagrams.

6.4 Types of capacitor

The value of a capacitor is usually printed on it, or it is marked with a set of coloured bands – see below. Also marked on the capacitor is its maximum safe working voltage (e.g. 100 V). If this voltage is exceeded, the capacitor would be damaged by the flow of current through the insulating dielectric. There are many different types of capacitor, and each one is generally distinguished by the type of dielectric used in its construction. Fig. 6.4 shows five different types of capacitor in general use.

An **electrolytic capacitor** has a value in the range 1 μF to 50 000 μF. It generally has a 'Swiss roll' construction in which the dielectric is a very thin layer of metal oxide between electrodes of aluminium or tantalum foil. The foil is rolled up to obtain a larger area of electrodes in a small volume. A tantalum capacitor offers a high capacitance in a small volume, and some of them are colour-coded as shown in Fig. 6.5. Electrolytic capacitors are not suitable for frequencies above about 10 kHz and are commonly used in timers (Section 6.6) and as smoothing capacitors in power supplies (Chapter 7).

A polyester capacitor is an example of a **plastic-film capacitor**. Polypropylene, polycarbonate and polystyrene capacitors are also types of plastic-film capacitor. The plastic, with the exception of

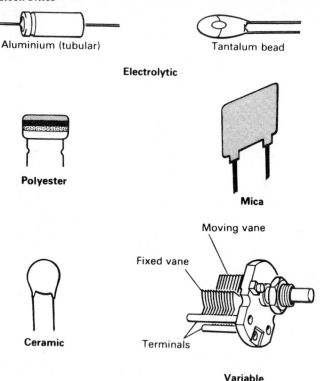

Fig. 6.4 Five types of capacitor

polystyrene which has a low melting point, has a metallic film deposited on it by a vacuum evaporation process – these are called metallised capacitors. Polyester capacitors can have values up to 10 μF, but other plastic-film capacitors have lower values. Working voltages of polyester capacitors can be as high as 400 V, and of polycarbonate capacitors, 1000 V. The Mullard C280 series of polyester capacitors are colour-coded as shown in Fig. 6.5.

Mica capacitors are rather more expensive than plastic-film capacitors and they are made by depositing a thin layer of silver on each side of a thin sheet of mica. These are excellent capacitors for use at high frequencies. They have values in the range 1 pF to 0.01 μF, have excellent stability and are accurate to ±1% of the marked value.

Ceramic capacitors consist of a silver-plated ceramic tube or disc and are excellent for use at high frequency.

Variable capacitors have low values, e.g. 500 pF, and are con-

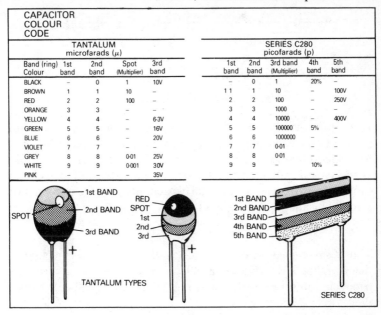

| CAPACITOR COLOUR CODE | | | | | | | | | |
| TANTALUM microfarads (μ) | | | | SERIES C280 picofarads (p) | | | | | |
Band (ring) Colour	1st band	2nd band	Spot (Multiplier)	3rd band	1st band	2nd band	3rd band (Multiplier)	4th band	5th band
BLACK	–	0	1	10V	–	0	1	20%	–
BROWN	1	1	10	–	1	1	10	–	100V
RED	2	2	100	–	2	2	100	–	250V
ORANGE	3	3	–	–	3	3	1000	–	–
YELLOW	4	4	–	6·3V	4	4	10000	–	400V
GREEN	5	5	–	16V	5	5	100000	5%	–
BLUE	6	6	–	20V	6	6	1000000	–	–
VIOLET	7	7	–	–	7	7	0·01	–	–
GREY	8	8	0·01	25V	8	8	0·01	–	–
WHITE	9	9	0·001	30V	9	9	–	10%	–
PINK	–	–	–	35V	–	–	–	–	–

SPOT — 1st BAND — 2nd BAND — 3rd BAND

RED SPOT — 1st — 2nd — 3rd

TANTALUM TYPES

1st BAND — 2nd BAND — 3rd BAND — 4th BAND — 5th BAND

SERIES C280

Fig. 6.5 Colour coding of tantalum and the C280 series of polyester capacitors

structed so that one metal plate moves relative to a fixed metal plate. The plates are separated by air or plastic sheet which acts as the dielectric. Variable capacitors are often used in radio receivers for tuning in to different stations.

6.5 Combinations of capacitors

The capacitors described above fall into two main categories: polarised and unpolarised. Aluminium and tantalum capacitors are polarised, and polyester, polystyrene and ceramic capacitors are unpolarised. The symbols for these two types of capacitor are shown in Fig. 6.6. The '+' sign on one of the terminals of the polarised capacitor indicates that this capacitor must be connected the right way round in a d.c. circuit. Note that the two parallel lines of the symbol separated by a space (filled with a dielectric) are indicative of the construction of a capacitor.

If the area of each plate of a capacitor is doubled, their separation remaining the same, the capacitance of the capacitor is doubled.

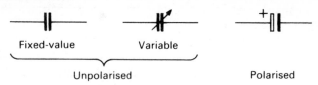

Fig. 6.6   Two symbols for capacitors

Also, if the separation of the plates is halved, their area remaining the same, the capacitance is also doubled. These relationships give us a clue to the capacitance of two capacitors connected in parallel as shown in Fig. 6.7a. The total area of the plates is effectively the sum of the two areas, so the total capacitance C of the combination is found by adding their values together. Thus

$$C = C_1 + C_2 \text{ (parallel)}$$

Thus if two capacitors of 10 µF and 50 µF are connected in parallel, the combined capacitance is 60 µF. But if two capacitors are connected in series as shown in Fig. 6.7b, the formula for finding their total capacitance is

$$C = \frac{C_1 \times C_2}{C_1 + C_2} \text{ (series)}$$

i.e. it is equal to the product of their values divided by the sum of their values. Thus if the above two capacitors of 10 µF and 50 µF are connected in series, the total capacitance of the combination is given by

$$C = \frac{10 \times 50}{10 + 50} = \frac{500}{60} = 8.33 \, \mu\text{F}$$

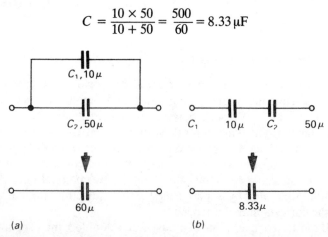

Fig. 6.7   Two capacitors connected in (a) parallel, and (b) series

The combined capacitance of three capacitors connected in series is given by the equation

$$\frac{1}{C} = \frac{1}{C_1} + \frac{1}{C_2} + \frac{1}{C_3}$$

Note that when two or more capacitors are connected in parallel, their combined capacitance is more than the capacitance of the larger value. And when they connected in series, their combined capacitance is less than the smaller value capacitor. You should compare these equations with the equations for series and parallel resistor combinations in Section 5.2.

6.6 Charging and discharging capacitors

Most timers and oscillators are based on the simple circuit shown in Fig. 6.8. Here a capacitor, C_1, and a resistor, R_1, are connected in series with a battery of e.m.f. E. On closing the switch, SW_1, the p.d. across the capacitor rises as shown by the graph. The p.d. rises relatively fast at first and then more slowly as the p.d. approaches the e.m.f. E of the battery. It is this delay in the charging (and discharging) of a capacitor which enables timers and oscillators to be designed. The time taken for the p.d. across the capacitor to rise to two-thirds of E is known as the **time constant** of the capacitor/resistor (RC) circuit. The time constant T is dependent on the values of both R_1 and C_1 and is given by the simple equation

$$T = R_1 \times C_1$$

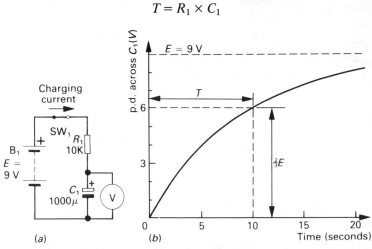

Fig. 6.8   The time constant of a CR combination

This equation gives the time constant in seconds if C_1 has a value in farads and R_1 has a value in ohms. For example, suppose the two values are $C_1 = 1000\ \mu F$ and $R_1 = 10\ k\Omega$. Thus $C_1 = 1000 \times 10^{-6} = 10^{-3}\ F$ and $R_1 = 10 \times 10^3 = 10^4\ \Omega$. Therefore $T = 10^{-3}\ F \times 10^4 \Omega = 10\ s$. This means that if $E = 9\ V$, it takes 10 seconds for the p.d. across the capacitor to rise from 0 V to two-thirds of E, i.e. 6 V. In practical timer and oscillator circuits, the time constant is a convenient way of estimating the rate of charge of a capacitor. Note that, to be completely accurate, the p.d. should be measured rising to 63% of E. However, 'two-thirds' (i.e. 67%) is accurate enough for most practical circuits since many electrolytic capacitors can have a tolerance of $\pm 50\%$ or more.

Just to complete our examination of the RC circuit, suppose the capacitor is discharged once it has been charged. Fig. 6.9 shows what happens if the battery is isolated and the capacitor is allowed to discharge through resistor R_1. The graph shows that the p.d. across the capacitor falls fast at first and then more slowly as the p.d. reaches 0 V. Incidentally, the graphs for the charge and discharge of a capacitor are known, mathematically, as exponential curves. We do not need to look at the equation for this graph and how it gives rise to the time constant, but the charge and discharge graphs do have an interesting property. Fig. 6.10 shows the charging curve.

Suppose we carried on timing the charging of the capacitor after it had reached two-thirds of E, i.e. 6 V. We should find that after another time constant of 10 s, the p.d. across the capacitor would have risen by two-thirds of the remaining p.d. Since the remaining

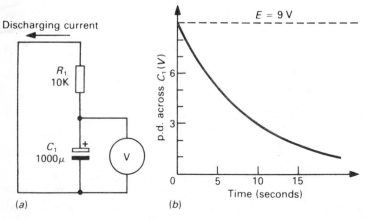

Fig. 6.9 Discharging a capacitor

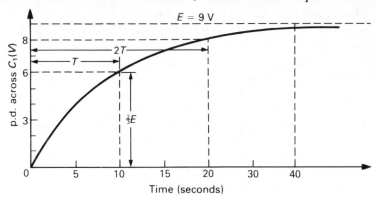

Fig. 6.10 The effects of successive time constants

p.d. is 3 V (9 V − 6 V), the p.d. would rise by a further 2 V to 8 V. After the next time constant of 10 s, the p.d. across the capacitor would have risen to two-thirds of 1 V (9 V − 8 V), i.e. by 0.67 V to 8.67 V. And so on. You can see that after three time constants, i.e. after 30 s in our example, the capacitor is almost fully charged, the p.d. across it having risen to 8.67 V, only 0.33 V short of its final p.d.

6.7 A practical timer

Fig. 6.1 showed the function of a timer. A timer is 'triggered' by an input signal and produces an output signal which lasts for a time T seconds. Now in order to make this function possible we need to connect together three black boxes as shown in Fig. 6.11:

black box 1 is the RC combination discussed above;

black box 2 is a device that detects when the p.d. across the capacitor has reached a certain value;

black box 3 is an output circuit that indicates that timing is in progress. The RC combination discussed above makes this function possible.

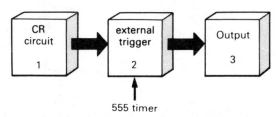

Fig. 6.11 The electronic system of a timer

Now black box 2 must be a device which

(*a*) makes the output signal go HIGH when it receives a trigger signal, and

(*b*) returns the output signal to LOW when it detects that the p.d. across the capacitor in the RC combination has risen by a certain amount.

Nowadays, a circuit designer would choose an integrated circuit (IC) to carry out the function of black box 2. And the most popular of the ICs available for this job is the device designated by manufacturers as the '555 timer' or 'triple-5 timer'. It is shown in Fig. 6.12, and comprises a small black plastic package having eight terminal pins for connecting it into a circuit. Inside the package is a silicon chip (Chapter 13) about 2 mm by 2 mm in area.

Now we do not need to know anything about the workings of this chip to use it as black box 2 in the timer system. Fig. 6.13 shows the circuit equivalent of the three interconnected black boxes. The main features of this circuit are as follows.

On closing switch SW_1 to switch on the power supply, the output signal is LOW, and the lamp L_1 is off. At this point the capacitor C_1 is discharged (by a transistor on the 555 timer chip). The timer is triggered by momentarily pressing SW_2, a push-to-make, release-to-break switch. This opens the internal switch of the 555 timer and at the same time makes the output of the 555 timer go HIGH. The lamp lights and the capacitor C_1 begins to charge through resistor R_1. The lamp remains lit until the p.d. across the capacitor reaches two-thirds (exactly) of the e.m.f. of the supply. Pin 6 of the 555

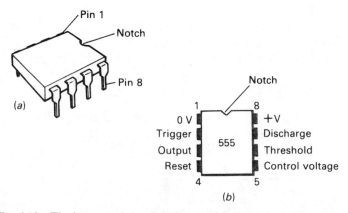

Fig. 6.12   The integrated circuit 555 timer: (a) what it looks like, and (b) the identity of its pins

senses when the p.d. across the capacitor reaches 6 V (two-thirds of 9 V). At this instant, the internal switch on the 555 closes, C_1 is instantly discharged via pin 7, and the output of the 555 goes LOW. The lamp switches off and the circuit now waits for another trigger signal.

The time that the lamp remains lit is given by the equation $T = 1.1 \times C_1 \times R_1$. So, using the values in Fig. 6.13,

$$T = 1.1 \times (1000 \times 10^{-6}\,\text{F}) \times (100 \times 10^3\,\Omega)$$
$$= 110\,\text{s}$$

This calculation assumes that the 1000 μF capacitor actually has the value marked on it, but we know that electrolytic capacitors have big tolerances. A second problem with large-value electrolytic capacitors is that some of the charging current leaks through the capacitor, from one plate to the other. These practical problems generally make the time delay longer than calculated. So it is often best to replace R_1 by a 100 kΩ variable resistor and adjust the time delay to precisely 100 s. But once set, the 555 timer will deliver the same delay every time it is triggered, even though the e.m.f. of the power supply may fall with time. This simple timer system is capable of giving reliable timing up to about one hour.

Finally, note that if you want the timer to control more power,

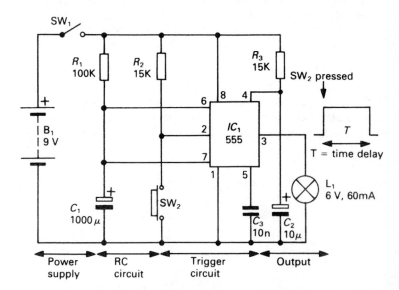

Fig. 6.13 A timer based on the 555 IC

then a relay (Section 3.4) should replace the lamp. The contacts on the relay should be rated to switch the current required by the high power load. Incidentally, the function of resistor R_2 is to make sure that the trigger input (pin 2) is held HIGH before and after switch SW_1 is operated. And components, R_3 and C_2, connected to pin 4 ensure that the timer does not begin to operate when the power supply is switched on. In fact, pin 4 could be used to reset the output to LOW and turn the lamp off, at any time during a delay. Capacitor C_3 ensures reliable operation of the timer in an electrically noisy environment.

Note that this timer is sometimes called a **monostable**, the word meaning 'one stable state', i.e. the output waveform is stable only when it is LOW. The HIGH state during the time delay is a temporary and unstable state of the timer.

6.8 A practical oscillator

As explained in Section 6.1, the waveform produced by a rectangular wave oscillator comprises a continuous series of HIGH and LOW signals. Fig. 6.14 shows the systems diagram for a simple rectangular wave oscillator. It is based on three black boxes:

black box 1 is the RC combination described above;

black box 2 is a device that detects when the p.d. across the capacitor has reached a charged and a discharged value;

black box 3 is an output circuit that makes use of the HIGH/LOW output signals.

Now the 555 timer IC can be made to operate as an oscillator as well as a timer. Fig. 6.15 shows a circuit which performs the function of the three interconnected black boxes. The circuit is slightly more complicated than the timer circuit of Fig. 6.13. Note that a resistor R_2 is now connected between pins 6 and 7 of the IC. And pin 4, the

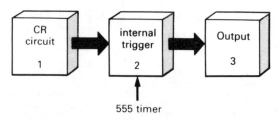

Fig. 6.14 The electronic system of an oscillator

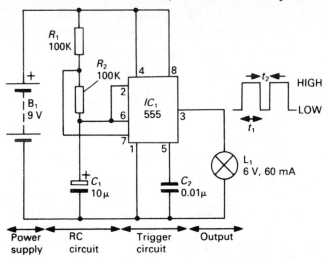

Fig. 6.15 An oscillator based on the 555 timer IC

reset pin, is connected directly to the power supply. The 555 timer takes the part of black box 2 in the systems diagram of Fig. 6.14. The circuit behaves as follows.

On switching on the power supply, the p.d. across the capacitor C_1 is 0 V, and the internal transistor switch is open. At this point, the output waveform is HIGH and the lamp L_1 is lit. The p.d. across C_1 immediately begins to rise as it charges through resistors R_1 and R_2. At the instant that the p.d. across C_1 reaches exactly two-thirds of the e.m.f. of the supply, i.e. 6 V in this case, the internal transistor switch closes and C_1 discharges via resistor R_2 and pin 7 of the 555 timer. Simultaneously, the output wave form goes LOW, and L_1 goes out. The output waveform remains LOW as the p.d. across C_1 falls. Pin 6 of the 555 timer senses when the p.d. has fallen to one-third of the supply e.m.f., i.e. 3 V in this case. At this instant, the internal transistor switch opens, the output waveform goes HIGH once again, L_1 lights, and the capacitor charges once again through resistors R_1 and R_2. The cycle repeats, the capacitor alternately charging and discharging between 3 V and 6 V, and the lamp switching on and off as the waveform goes HIGH and LOW, respectively.

Note that C_1 charges through R_1 and R_2 and discharges through R_2. This is reflected in the two equations for the HIGH (C_1 charging through R_1 and R_2), and LOW (C_1 discharging through R_2) times of the output waveform from the oscillator. These times are as follows:

$$\text{HIGH time, } t_1 = 0.7 \times (R_1 + R_2) \times C_1$$
$$\text{LOW time, } t_2 = 0.7 \times R_2 \times C_1$$

Thus, using the values of C_1, R_1 and R_2 in Fig. 6.15, the t_1 and t_2 times are as follows:

$$t_1 = 0.7 \times (200 \times 10^3) \times 10 \times 10^{-6} = 1.4 \text{ s}$$
$$t_2 = 0.7 \times (100 \times 10^3) \times 10 \times 10^{-6} = 0.7 \text{ s}$$

The lamp flashes on for 1.4 s and off for 0.7 s. In practice, these times might be different from the calculated values due to the tolerance of the components used. But if precise HIGH/LOW times are required, it is easy to replace R_1 or R_2, or both, by variable resistors. Note that it is impossible with this simple circuit to get the HIGH and LOW times equal, i.e. to produce a waveform which is a square wave. If C_1 has a value of 100 μF, R_1 and R_2 remaining the same, the HIGH and LOW times each become ten times longer, i.e. 14 s and 7 s. If the lamp is replaced by a relay, the relay contacts could be used to operate higher-power lamps, motors, etc. (If you are puzzled about the factor '0.7' in the above equations, don't be: in the proof of these equations for the astable function of the 555, a factor '0.693' appears and we have approximated this to '0.7'.)

Note that this oscillator is sometimes called an **astable**, the word meaning 'no stable state', i.e. the output is not stable in either of the two states, HIGH or LOW. It oscillates from one state to the other.

7

Rectification and Diodes

7.1 What rectification is

Fig. 7.1 shows that rectification is the conversion of an a.c. waveform into a d.c. waveform. This is a useful function and is the basis of mains-operated power supplies that provide the d.c. voltages required, for example, by video recorders, hi-fi systems and microcomputers. In a practical a.c.-to-d.c. power supply, the varying a.c. waveform shown in Fig. 7.1 has to be 'smoothed' by capacitors before it is an acceptable d.c. supply. The conversion of an a.c. voltage to d.c. voltage is used in an amplitude-modulated (AM) radio receiver for detecting the radio signal captured by its aerial, as explained in Chapter 16.

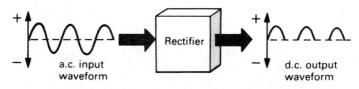

Fig. 7.1   Rectification converts a.c. to d.c.

Rectification involves the use of one or more devices called **diodes** or simply **rectifiers**. These are devices that act as one-way valves for the passage of electrons. Indeed, the first diodes were thermionic valves (Section 1.9) which were so named because they allow electrons to pass easily through them in one direction. The symbol for a diode is shown in Fig. 7.2a; Fig. 7.2b shows two types of diode, one that allows a maximum current of 1 A to flow through it and the other a maximum of 13 A. Note that the arrow of the diode symbol

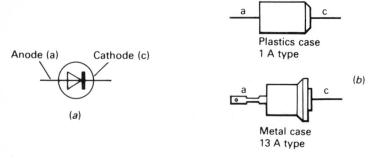

Fig. 7.2   Rectifier diodes: (a) general symbol; and (b) the appearance of two types

points in the direction of conventional current flow through it, so electrons flow through the diode against the direction of this arrow. Conventional current flows through a diode from its anode terminal to its cathode terminal. On plastics diodes, the cathode terminal is often marked with a red, black or white band. When the anode of a diode has a more positive voltage than the cathode, the diode is said to be 'forward-biased' and current flows easily through it (Fig. 7.3a). When the anode is more negative than the cathode, it is said to be 'reverse-biased' and no current flows through it (Fig. 7.3b).

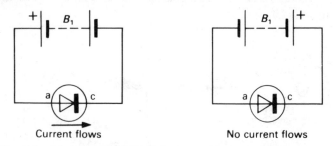

Fig. 7.3   (a) A forward-biased diode, and (b) a reverse-biased diode

7.2 The p–n junction

The rectifying action of a semiconductor diode is due to the electrical effects that occur at the junction between an n-type semiconductor and a p-type semiconductor, a region known as a **p–n junction**. Fig. 7.4a shows that an interesting thing happens when this p–n junction is made: electrons flow across the junction from the n-type material to the p-type material, and holes flow in the

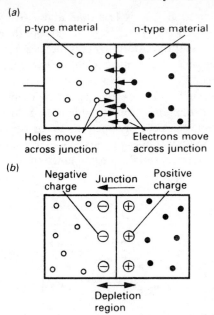

Fig. 7.4 (a) Charge flow across an unbiased p–n junction, and (b) the formation of a depletion region

opposite direction. The reason for this flow of charge should be clear from what has been said (Section 2.5) about the properties of p-type and n-type semiconductors. Electrons move to fill holes and holes move to capture electrons, a process called diffusion. (It is diffusion that is responsible for the way the aroma of perfume and aftershave reaches people nearby.)

The movement of electrons and holes across the junction depletes the n-type material of holes and the p-type material of electrons. A narrow 'depletion region' is created that is less than one millionth of a metre wide, as shown in Fig. 7.4b. The resulting positive charge in the n-type material and negative charge in the p-type material produce what is known as a **potential barrier** that opposes the further diffusion of charge across the junction. The diode's potential barrier (an internal property of the p–n junction) is increased or decreased depending on the polarity of an external p.d.

Suppose an external p.d. makes the p-type material more positive than the n-type material as shown in Fig. 7.5a. The effect is to decrease the potential barrier and reduce the width of the depletion region. If the external p.d. is less than about 0.5 V (for a p–n junction based on silicon), no current flows from the p-type to the n-type material. But if the external p.d. is increased to about 0.6 V, the depletion region vanishes and current flows across this forward-biased p-n junction. If the external p.d. is applied to the p–n junction as shown in Fig. 7.5b, electrons and holes move away from the junction, the depletion region widens and no current flows across this reverse-biased p–n junction. In an actual diode, the anode terminal is connected to the p-type material and the cathode terminal to the n-type material. The p–n junction acts as a 'one-way valve' for the flow of electricity.

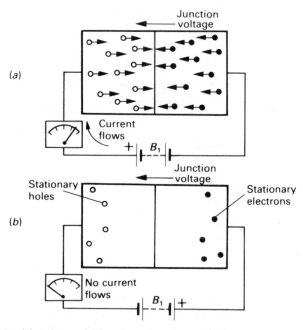

Fig. 7.5   (a) A forward-biased p–n junction, and (b) a reverse-biased p–n junction

If measurements are made of the current flowing through a silicon diode for different values of the p.d. across it, a graph similar to that shown in Fig. 7.6 is obtained. The graph is known as the current–voltage characteristic of a p–n silicon diode. It is clear from this graph that the current flowing across the p–n junction increases sharply as the external voltage changes from about 0.5 V to 0.7 V. This means that the resistance of the p–n junction falls sharply over

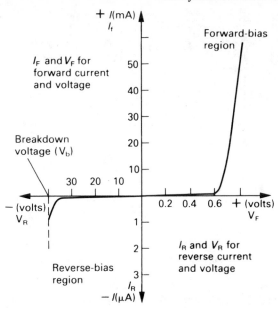

Fig. 7.6 The voltage/current characteristic of a silicon junction diode

this voltage range. In the reverse-bias region, no current flows through the diode until the voltage reaches the reverse breakdown voltage, V_b. This voltage varies from a few volts to a few hundred volts depending on the type of diode used. The diode is usually ruined if V_b is exceeded.

7.3 Using diodes in power supplies

An oscilloscope is used in the circuit of Fig. 7.7 to see how a diode changes a.c. to d.c. A 50 Hz a.c. supply of a few volts is connected across diode D_1 which is in series with resistor R_1. The oscilloscope monitors the changes of voltage across resistor R_1. With SW_1 closed to short-circuit D_1, the oscilloscope displays an a.c. waveform, as might be expected since it is effectively connected across the a.c. supply. When SW_1 is opened, the trace shows that one half of the a.c. waveform is removed by the diode. This happens because D_1 is alternately forward-biased and reverse-biased by the a.c. supply. The diode allows current to flow through it and R_1 only when terminal T_1 of the supply is positive with respect to terminal T_2. The current through R_1 is a varying d.c. and the circuit is called a

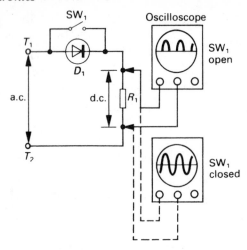

Fig. 7.7   Half-wave rectification

half-wave rectifier. Note that the half-wave rectifier loses one half of the a.c. waveform.

A better rectifier circuit is shown in Fig. 7.8; this is known as a **full-wave rectifier**. The 'lost' half wave now contributes to the d.c. output of the rectifier circuit since current flows through resistor R_1 in the same direction for both halves of the a.c. waveform. Thus when terminal T_1 of the a.c. supply is positive with respect to T_2, current flows through R_1 in the direction shown by the full-line arrows. And when T_2 is positive with respect to T_1, the direction of

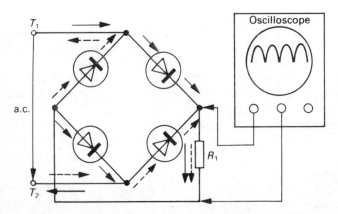

Fig. 7.8   Full-wave rectification

the current through R_1 is unaltered and is represented by the dotted-line arrows.

The outputs of the half-wave and full-wave rectifier circuits are varying d.c. and need to be smoothed to produce a steady d.c. supply. Smoothing is achieved with a capacitor, C_1 as shown in Fig. 7.9 for the half-wave rectifier. When terminal T_1 is positive, the diode conducts and C_1 charges up to the peak voltage of the a.c. supply. During this time current is also flowing through R_1. When the voltage at T_1 begins to fall during the first half cycle of the supply, current continues to flow through R_1 and C_1 discharges through it. If the time constant, $T = C_1 \times R_1$ (Section 6.5), is long enough, C_1 discharges little during the second half cycle when D_1 is reverse-biased. The effect is a smoothed d.c. supply as shown by the graphs.

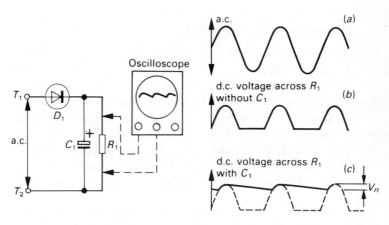

Fig. 7.9 Using a smoothing capacitor. The graphs show: (a) unrectified a.c.; (b) half-wave rectified d.c.; (c) smoothed a.c.

The d.c. output voltage across R_1 has a small ripple voltage superimposed on it caused by the small fall in voltage across C_1 during the second half cycle. For good smoothing action, the product $R_1 \times C_1$ must be larger than 1/50 second (0.02 s), i.e. the time between peak voltages across R_1 for a 50 Hz a.c. supply. For efficient smoothing of the d.c. output, the product $R_1 \times C_1$ ought to be about five times larger than 0.02 s. Thus if $R = 1$ kΩ, $C_1 \times R_1 = 0.1$ s and

$$C = (0.1 \text{ s}/R_1) = (0.1/10^3) = 10^{-4} \text{ F} = 100 \text{ μF}$$

In practical rectifier circuits, smoothing capacitors are of the electrolytic variety and have values between about 470 μF and 2200 μF. Perhaps you can see why a capacitor of lower value would suffice to smooth the d.c. output of a full-wave rectifier?

In practical circuits it is important to use a diode that will not be damaged by the current flowing through it. This current is called the **maximum forward current** and it is usually indicated by the symbol I_F. Also note that since the capacitor discharges very little during the half cycle when the voltage at T_1 changes from its maximum positive to its maximum negative value, the p.d. across the diode reaches almost twice the peak voltage of the a.c. supply. The diode has to be rated to withstand this **maximum reverse voltage** (V_R) without being damaged. Thus the 1N4001 has an I_F value of 1 A, and a V_R value of 50 V; and the 1N4006 diode has an I_F value of 1 A and a V_R value of 800 V.

7.4 Zener diodes and voltage regulators

A zener diode is a special type of diode used in power supplies to stabilise a d.c. voltage. Like all diodes, a zener diode has an anode terminal and a cathode terminal. But the characteristic 'knee' in the line at the cathode terminal in the symbol (Fig. 7.10a) indicates its special function. The voltage–current characteristic of a typical zener diode shown in Fig. 7.11 reflects this 'knee'. The graph shows that the zener diode is operated in the reverse-bias breakdown region where a normal rectifier diode would be damaged. In this region, large current changes through it produce very little voltage change across it, i.e. the slope of the characteristic is very steep. To maintain the zener diode in the reverse-breakdown region, the current through it must not fall below about 5 mA (for this diode).

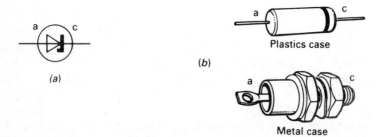

Fig. 7.10   (a) The symbol for a zener diode, and (b) the appearance of two types of zener diode

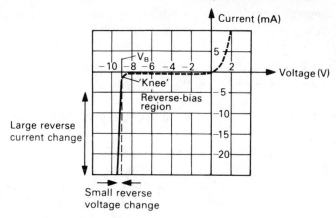

Fig. 7.11 The voltage–current characteristic of a zener diode

The reverse breakdown voltage, V_B, of a zener diode is usually marked on it. Thus if '5V6' is printed on a zener diode, it will stabilise a voltage of 5.6 V. The values of V_B follow a series of preferred values (just as for resistor values – see Section 3.3). These values are 2V7, 3V, 3V3, 3V9, 4V3, 4V7, 5V1, 5V6, 6V2, 6V8, 7V5, 8V2, 9V1, 10V, and so on.

A rudimentary voltage stabiliser using a zener diode is shown in Fig. 7.12. This circuit stabilises the output voltage at 5.6 V for a wide range of input voltages. There are three main factors to consider in this simple design: the minimum input voltage for which the output voltage remains constant at 5.6 V; the power dissipation of the diode at the maximum input voltage; and the maximum current that can be drawn by the load resistor R_L to maintain voltage stabilisation for a specified input voltage. Let's consider each requirement in turn, assuming that the diode has the following ratings: $V_B = 5.6$ V;

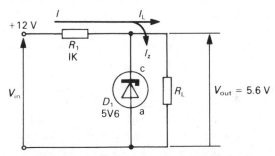

Fig. 7.12 A simple voltage-regulated supply

maximum power dissipation = 500 mW; minimum reverse current = 5 mA. The BZY88 series of zener diodes have a power rating of 500 mW.

1 **Minimum input voltage**   Suppose the load current, I_L, is zero, i.e. $I = I_Z$. First calculate the p.d. across R_1. If $I_Z = 5$ mA and $R_1 = 1$ kΩ, the p.d. across $R = 5$ mA $\times$ 1 kΩ = 5 V. Thus the minimum input voltage is the sum of the p.d. across R_1 and the p.d. across D_1, i.e. $V_{in} = 10.6$ V.

2 **Power dissipation in the diode**   Suppose the maximum input voltage is 20 V. This input voltage makes the p.d. across $R_1 = 20 - 5.6 = 14.4$ V. Thus the current through $R_1 = 14.4$ V/1 kΩ = 14.4 mA = I_Z. The power dissipation, P, in the diode is given by

$$P = I_Z \times V_Z = 14.4 \text{ mA} \times 5.6 \text{ V} = 80.6 \text{ mW}$$

which is well within the maximum rating of 500 mW.

3 **Maximum load current**   Suppose $V_{in} = 12$ V. Now p.d. across $R_1 = 12 - 5.6 = 6.4$ V. Current through $R_1 = 6.4$ V/1 kΩ = 6.4 mA. If at least 5 mA of this has to flow through D_1 to maintain it in the breakdown region, the maximum load current is 1.4 mA which is preciously small! Robbing the diode of more current than this would take the operating point of the diode below the 'knee' of its characteristic and voltage stability would be lost.

Of course, you could allow more load current to be drawn from this simple stabiliser by using either a higher power (and more expensive) zener diode (e.g. the 5 W 1N5333 series), or by reducing the value of the series resistor R_1 (provided the maximum power rating of the diode is not exceeded). Fortunately, the more elegant solution shown in Fig. 7.13 is to hand. This is a practical circuit that provides a nominal 9 V, 2 A stabilised d.c. supply from the a.c. mains. The circuit uses an npn transistor (Chapter 8) to increase the available output current of the voltage stabiliser without extra demands on the diode. The general operation of this power supply can be understood by noting that it is made up of four building blocks.

1 Building block 1 is a step-down mains transformer T_1 that produces a 12 V a.c. output from the 240 V a.c. mains supply.
2 Building block 2 is a bridge rectifier Br_1 that contains four silicon rectifier diodes and produces a full-wave rectified d.c. output from the 12 V a.c. supply.
3 Building block 3 comprises a smoothing capacitor C_1 that pro-

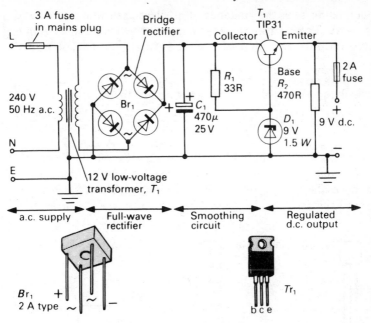

Fig. 7.13 A mains-operated d.c. supply

duces a smoothed d.c. output of about 14 V. This voltage will vary with variations in the mains voltage.

4 Building block 4 is based on a zener diode D_1 and npn transistor Tr_1.

Now the operation of the transistor is described in Chapter 8. Its main purpose here is to provide a higher current than is possible with the simple circuit shown in Fig. 7.12. Indeed the output current is increased by the current gain of the transistor. Resistor R_1 continues to provide the small reverse current for the diode plus that required by the base of the transistor. The zener diode D_1 holds the voltage at the base terminal of Tr_1 at 10 V. The voltage at the emitter of the transistor is 0.6 V less than this, i.e. 9.4 V. The amplified current flowing between the collector and emitter terminals of the transistor is the current drawn by the load, and this is limited to 2 A by the fuse. In order to protect the transistor from damage due to the heat generated within it, it is usual to mount it on a 'heat sink'. The inset pictures show the appearance and pin connections of the bridge rectifier and transistor. Remember that if you are 'practically-minded' and want to construct this power supply, do seek the guidance of an electrician who can show you how to 'earth' it properly. For example, note that the metal frame of the transformer is connected to the earth (E) terminal of the mains supply.

Nowadays circuit designers tend to use purpose-designed voltage regulators that contain in a single package the zener diode, associated resistors and current-amplifying transistors. Two typical voltage regulators are shown in Fig. 7.14a. These replace the zener diode, resistor R_1 and transistor in Fig. 7.13. A 12 V, 2 A voltage regulator such as this has internal circuitry that automatically limits the output current to 2 A maximum (called 'current limiting'), and reduces the current passing through it if it overheats (called 'thermal shutdown').

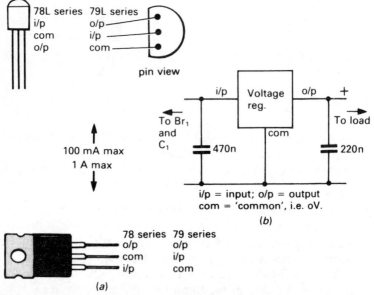

Fig. 7.14　(a) Two examples of packaged voltage regulators; (b) the way a voltage regulator is used

7.5 Light-emitting diodes

By careful selection of p-type and n-type semiconductors it is possible to make a p–n junction which emits light when it is forward-biased. This type of diode is called a light-emitting diode (LED). The shape of a typical LED and its circuit symbol are shown in Fig. 7.15. In one common type of LED, the cathode is identified

Fig. 7.15 (a) The shape, and (b) the symbol of a typical LED

by a 'flat' where the leads enter the LED. It is only necessary for a current of a few milliamperes to pass through the p–n junction to make the LED emit light. Thus LEDs are ideal as indicators in all types of application, especially for battery-powered devices where a filament lamp would absorb too much power. Groups of LEDs are also used in seven-segment displays and bargraph displays as explained in Section 11.2.

When a p–n junction is forward-biased, electrons move from the n-type to the p-type material where they combine with holes quite close to the junction. Holes also move from the p-type material to combine with electrons in the n-type material. In a p–n junction based on silicon, the combination of electrons and holes releases energy as heat which simply warms up the junction. However, if the p–n junction is made from the semiconductor gallium arsenide, energy escapes from the junction in a fairly narrow band at infrared wavelengths. And by mixing other substances in very small quantities with the gallium arsenide, visible light is produced which escapes from the surface of the diode. The table below shows the wavelengths of the light emitted by different types of LEDs.

Semiconductor	Light	Wavelength (µm)
Gallium arsenide (GaAs)	infrared	0.9
Gallium arsenide phosphide (GaAsP)	red	0.65
Gallium phosphide (GaP)	green	0.56
Gallium indium phosphide (GaInP)	yellow	0.50

Gallium arsenide offers some advantages over silicon, and its properties are described in Section 13.10. The wavelengths of different radiations in the electromagnetic spectrum are listed in Fig. 16.2, and Section 16.9 explains how infrared emitting LEDs are used in optical communications systems.

An LED emits light if the p.d. across it is about 2 V. The table below shows the value of this forward-bias voltage, the maximum allowable reverse-bias voltage and forward-bias current for three different LEDs.

LED	Forward-bias voltage at 10 mA/V	Maximum reverse-bias voltage (V)	Maximum forward-bias current (mA)
Red	2.0	5	40
Green	2.2	5	40
Yellow	2.4	5	40

A resistor must be connected in series with an LED if it is to be lit by a p.d. greater than 2 V. Suppose the LED is to be lit by a 9 V battery as shown in Fig. 7.16. The value of resistor R_1 is calculated as follows. Assume that the current through the LED is 10 mA and the p.d. across it is 2 V.

p.d. across $R_1 = (9\ V - 2\ V) = 7\ V$

current through $R_1 = 10\ mA = 7\ V/R_1$

$R_1 = (7\ V/10\ mA) = 700\ \Omega$

Therefore a suitable value for R_1 is 680 Ω, but remember that an LED will light for currents in the range about 5 mA to 40 mA. Higher values of R_1 will reduce the current through the LED and it will not light as brightly, but it will last longer.

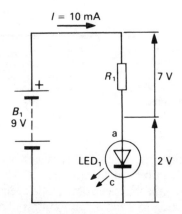

Fig. 7.16   Calculating the value of the series resistor for an LED

7.6 Power control with thyristors and triacs

These devices are used in power control circuits and, like rectifiers, depend on the properties of the p–n junction. They are to be found in foodmixers, electric drills and lamp dimmers. Fig. 7.17a shows the appearance of a typical **thyristor**. This device was formerly known as a *silicon controlled rectifier* (s.c.r.) since it is a rectifier that controls the power delivered to a load, e.g. a lamp or motor. The symbol for a thyristor is shown in Fig. 7.17b: it looks like a diode symbol with anode and cathode terminals, but with the addition of a third terminal called a gate, g. A simple demonstration of d.c. power control using the thyristor, Thy, is shown in Fig. 7.17c. This circuit forward-biases the thyristor but the thyristor does not conduct until a positive voltage is applied to the gate terminal by closing switch SW_1. This positive voltage allows a small current to flow into

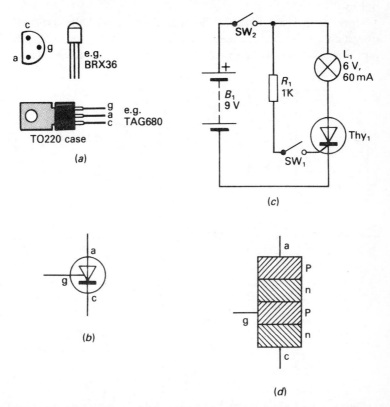

Fig. 7.17 The thyristor: (a) two types of package; (b) circuit symbol; (c) a simple power control circuit; (d) internal construction

the gate terminal and the thyristor 'fires'. However, conduction continues and the lamp remains lit even though the positive gate voltage is removed by opening SW_1. The only way to switch the thyristor off is to open switch SW_2. Four layers of semiconductors make up the thyristor's p-n-p-n sandwich construction as shown in Fig. 7.17d. Note that the word thyristor is derived from the Greek *thyra*, meaning door, and indicates that the thyristor is either open or closed.

It is possible to use a thyristor to control a.c. power by allowing current to be supplied to the load during only part of each cycle. Fig. 7.18 shows the basic circuit and waveforms. In a practical circuit, the gate pulses are applied automatically at a selected stage during part of each cycle. Thus half power to the load is achieved by applying the gate pulses at the peaks of the a.c. waveform. More or less power in the load is achieved by changing the timing of the gate pulses.

Since the thyristor switches off during the negative half cycles, it is only a half-wave device (like a rectifier) and allows control of only half the power available in a.c. circuits. A better device is a **triac**,

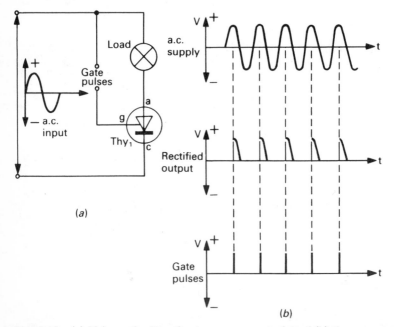

Fig. 7.18 (a) Using a thyristor for a.c. power control, and (b) the waveforms that make it work

which comprises two thyristors connected in parallel but in opposition and controlled by the same gate, i.e. it is bi-directional and allows current to flow through it in either direction. Fig. 7.19a shows its symbol. The terms anode and cathode have no meaning for a triac; instead the contact near to the gate is called 'main terminal 1' (MT1), and the other 'main terminal 2' (MT2). The gate trigger voltage is always referred to MT1, just as it is referred to cathode in the thyristor. As with the thyristor, a small gate current switches on a very much larger current between the main terminals. A typical

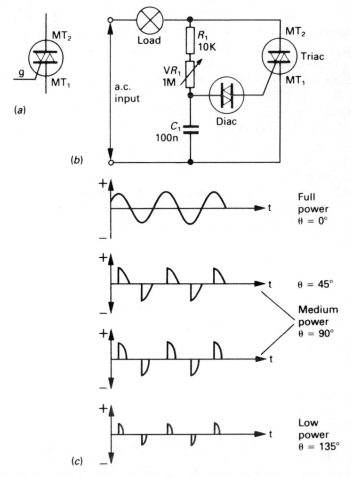

Fig. 7.19 The triac: (a) circuit symbol; (b) a basic circuit for a.c. power control; (c) the waveforms that make it work

gate current is of the order of 20 mA but it is adequate for triggering triacs of up to 25 A rating.

If the triac is to be used in a lamp dimmer unit or a motor speed controller, there has to be some means of varying the a.c. power passing through the load. Fig. 7.19b shows how this is achieved using an 'RC phase shifter'. VR_1 is the dimmer control. The device shown as a *diac* is in effect two zener diodes connected back to back. It conducts in either direction when the voltage across C_1 reaches the diac's breakdown voltage of about 30 V. The burst of current through the diac 'fires' the triac. As you know from Section 6.6, the rate at which C_1 charges depends on the value of C_1 and that of the series resistor, $VR_1(+R_1)$. The greater the value of VR_1, the more slowly the capacitor charges and the later in each half-cycle is the lamp turned on and the dimmer the lamp.

The waveforms in Fig. 7.19c show how the triac controls power by chopping off part of each half cycle. The amount chopped off is indicated by the phase shift, θ. If θ is 0, the triac conducts through-out the whole of each half-cycle of the a.c. waveform and the load is at full power. As θ increases from 0 to 180°, more and more of each half cycle is chopped off, and the dimmer the lamp becomes. Practical a.c. power control circuits using triacs refine the basic design in Fig. 7.19b to give better low-level power control, and a reduction of the radio interference that is generated by the rapid turn-on and turn-off of the triac. You might be interested to know that a *quadrac* combines the triac and diac in a single package.

8

Amplifiers and Transistors

8.1 What amplifiers do

Fig. 8.1 shows what an amplifier does. It increases the power of the signal passing through it. Thus a low power signal, P_{in}, enters from the left, energy is drawn from the power supply, and a signal of higher power, P_{out}, leaves from the right. Note that an amplifier cannot increase the power of a signal without power being drawn from somewhere else, i.e. from a battery or some other source of electrical energy. In systems diagrams, it is usual to represent an amplifier black box as a triangle. This is in effect an arrowhead which symbolises the direction in which the signal travels from the input to the output of the amplifier.

Now the ratio of the output power to the input power is the power gain, A, of an amplifier. Thus

$$\text{power gain} = \frac{\text{output power}}{\text{input power}}$$

or

$$A = \frac{P_{out}}{P_{in}}$$

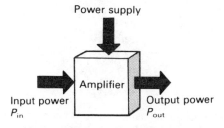

Fig. 8.1   The function of an amplifier

The power of electrical signals is measured in watts (Section 5.5). Suppose the output power of an amplifier is equal to 50 W. If the input power is equal to 0.01 W (10 mW), the power gain of the amplifier is given by

$$A = \frac{50 \text{ W}}{10 \text{ mW}} = 5000$$

Thus the output power from this amplifier is 5000 times bigger than the input power.

Amplifiers are very widely used in electronic systems, for example in hi-fi amplifiers to increase the power of audio frequency signals from tapes, records and compact disks before delivery to a loudspeaker. Video cassette recorders use amplifiers to amplify vision (video) signals. And communications satellites use radio frequency amplifiers to boost and retransmit signals sent up to them from ground stations. Let's take a closer look at the audio amplifier, which is perhaps the one you are most likely to come across.

8.2 Audio amplifiers

Audio amplifiers are to be found in hi-fi amplifiers (Fig. 8.2), car radios, walkie-talkies, sound and video cassette recorders, hearing

Fig. 8.2   A hi-fi stereo amplifier giving 70 W per channel and representing the state of the art in amplifier design
Courtesy: Quad Electronics

aids and satellite communications systems. In these systems their purpose is to increase the power of audio frequency (AF) signals in the range of human hearing, i.e. in the frequency range 20 Hz to 20 000 Hz (20 kHz). A 20 Hz AF signal sounds like a 'buzz', and one of 10 kHz like a high-pitched whistle. Few adults can hear AF signals above about 16 kHz, though many young children and animals do hear sounds that have frequencies higher than this. As we get older, our ears become increasingly less sensitive to audio frequencies over 10 kHz.

The easy way to amplify a weak audio signal is to use one of the integrated circuits designed for the job. Fig. 8.3 shows an audio IC designated TBA820M by its manufacturers. This device is housed in an 8-pin dual-in-line package which protects the 2 mm by 2 mm square silicon chip on which the amplifying circuit is made. When it is operated from a 12 V power supply, this device provides a maximum audio power of 2 W from an 8 ohm loudspeaker. It is actually designed to operate from any power supply voltage in the range 3 V to 16 V, though at 3 V its power output is much less than at 16 V. The manufacturers of this audio amplifier have also made sure that when it isn't actually amplifying a signal it draws very little current from the power supply – it is said to have a low 'quiescent current' (see Section 8.9 for more detail about this).

No integrated circuit audio amplifier contains all the components required to make a complete audio amplifier. If this were the case, the IC would only need four pins: two for the power supply and one each for the input and output signals. But as you can see from Fig. 8.3, pins are designated 'frequency compensation', 'bootstrap' and so on, and these labels indicate that components have to be con-

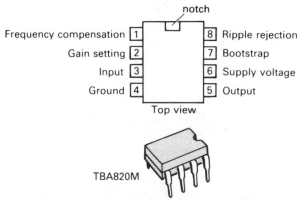

Fig. 8.3 The TBA820M audio amplifier

nected to these pins to ensure the proper functioning of this amplifier. A practical circuit that makes use of the TBA820M audio IC is described in Section 16.9.

8.3 The bandwidth of an amplifier

Most audio amplifiers are designed to amplify all frequencies within a specified band of frequencies by the same amount. Fig. 8.4 shows what this means, using a 'frequency response' graph. The level part of the graph is typical of the 'flat' frequency response of a hi-fi amplifier. However, the power gain of most audio amplifiers falls off sharply at frequencies below about 20 Hz and above about 20 kHz. Surprisingly, the power gain of some good quality audio amplifiers only begins to fall off at frequencies above 40 kHz. Manufacturers claim that this ability to amplify frequencies well above the range of human hearing improves the sound reproduction within the audio frequency range.

The frequency range over which the power gain does not fall below half the maximum power gain is called the **bandwidth** of the amplifier. Note that frequencies are plotted 'logarithmically' on the frequency response graph. This means that ten-fold increases of frequency occupy equal distances along the horizontal axis. Thus there is equal spacing between 10 Hz and 100 Hz, and between 100 Hz and 1000 Hz, and so on, so that the graph can accommodate the wide range of frequencies and show clearly the amplifier's frequency response in the 20 Hz to 2 kHz frequency range. If the calibrations along the frequency axis were linear, this 100-fold frequency range would occupy only about one-tenth of the full 20 kHz bandwidth. As it is, it occupies about one-half of the 20 kHz range.

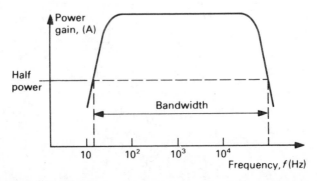

Fig. 8.4 The frequency response graph of an audio amplifier

Though a good quality amplifier has a flat frequency response over a wide range, it is common for manufacturers to provide a means to 'doctor' this response to suit individual tastes. This may take the form of a single potentiometer, a 'tone control', that can be adjusted to select relatively more high or low audio frequencies. Or the device might be equipped with a 'graphic equaliser' as shown in Fig. 8.5. This three-band graphic equaliser allows control of the frequency response in the low, middle and upper frequency range of the audio band. Hi-fi systems often have a ten-band graphic equaliser to allow more precise control of the frequency response.

Fig. 8.5　A 3-band graphic equaliser on a personal stereo cassette player
Courtesy: Aiwa Co Ltd

8.4 The decibel

The decibel (dB) is often used as a measure of the loudness of sounds, e.g. the sound from a low-flying jet at an airport might be 60 dB. It is also used when talking about the power gain of an amplifier. The decibel actually derives from the 'bel', which was introduced in the early days of telephone engineering to express how much more power one signal had than another. The unit is named in honour of Alexander Graham Bell, the pioneer of telephones. In practice, the bel turned out to be too large a unit so the decibel (one-tenth of a bel) is now used.

The reason that the bel is such as useful unit for comparing the strengths of audio frequency signals is that the response of the ear is logarithmic. This means that to the human ear a change in loudness is the same whether the sound power increases from 0.01 W to 0.1 W, or from 0.1 W to 1 W. Thus ten-fold increases in sound power seem like equal increases in loudness. Actually, the ear does not respond like this over the full range of audio frequencies, for it is more responsive to some audio frequencies than to others, being most sensitive at about 4 kHz.

Suppose that an amplifier increases the power of a signal from P_{in} to P_{out}. Then the power gain is the ratio P_{out}/P_{in}. The following equation expresses this gain in decibels.

$$\text{power gain} = 10 \times \log_{10}(P_{out}/P_{in})$$

For example, if the input power is 0.1 W and the output power is 10 W, the power gain is

$$
\begin{aligned}
& 10 \times \log_{10}(10\,\text{W}/0.1\,\text{W}) \\
= & 10 \times \log_{10}(100) \\
= & 10 \times 2 \\
= & 20\,\text{dB}
\end{aligned}
$$

Thus a ten-fold increase in power can be expressed as a 20 dB increase in power. A similar calculation shows that a 1000-fold increase in power is equivalent to a 30 dB increase; a million-fold increase to a 60 dB increase; and so on. Each ten-fold increase of power gain is equivalent to a 10 dB increase. Fig. 8.6 shows that

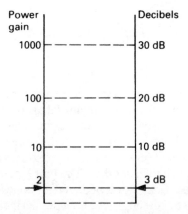

Fig. 8.6 A decibel scale for power gain

the decibel scale is a compact way of representing a wide range of power gains.

Manufacturers of audio amplifiers often use the decibel scale to specify the bandwidth of their amplifiers. For example, Fig. 8.7 is the frequency response graph of Fig. 8.4 redrawn to show the gain on the vertical axis in decibels. Now the bandwidth is shown as 20 Hz to 20 kHz between the 3 dB points. But why 3 dB?

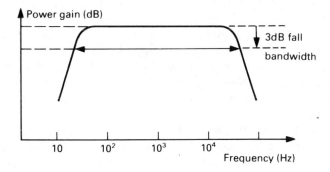

Fig. 8.7   The 3 dB points

Well, as you now know, the bandwidth of an amplifier is defined between the points on the frequency response graph where the power has fallen to half its 'flat' value. So the fall in gain expressed in decibels is $10 \times \log_{10}(1/2) = -3$ dB. The minus sign is important for it indicates a fall in power. In terms of the decibel, a doubling in power is equivalent to an increase of $+3$ dB, and a halving in power to a decrease of -3 dB.

Actually, it is easier to determine the gain of an amplifier by measuring the values of the p.d.s generated across input and output resistances by the input and output signals, rather than by measuring the input and output powers. In this case, the voltage gain of the amplifier is found from the following equation:

$$\text{voltage gain} = 20 \log_{10}(V_{\text{out}}/V_{\text{in}})$$

where V_{in} is the p.d. generated across the input resistance by the input signal, and V_{out} is the p.d. generated across the output resistance by the output signal. The equation assumes that the input and output resistances are the same. Thus if an amplifier boosts the strength of a signal from 10 mV to 10 V, it has a voltage gain of

$$20 \times \log_{10}(10 \, \text{V}/10 \, \text{mV})$$
$$= 20 \times \log_{10}(1000)$$
$$= 60 \, \text{dB}$$

and not 30 dB as it is for power gain.

8.5 Types of transistor

Invented in 1947 by the three-man team of Bardeen, Shockley and Brattain at the Bell Telephone Laboratories, USA, the transistor has become the most important basic building block of almost all circuits. Transistors were first used singly in circuits, but by the end of the 1960s silicon chips comprising several hundred transistors were being made. Nowadays, the most complex integrated circuits contain several hundreds or even thousands of transistors (see Chapter 13).

Transistors are still used as discrete, i.e. individual, components in circuits. Fig. 8.8 shows the distinguishing features of a small selection of transistors, which fall into two main categories depending on the way n-type and p-type semiconductors are used to make them. One sort is the **bipolar transistor** (the sort invented in 1947) of which there are two types, npn and pnp, as shown by the examples in Fig. 8.8a to 8.8c. The second sort is the unipolar transistor (a later invention) which is also called the **field-effect transistor** (FET). Examples of the FET, of which there are two sorts, n-channel and p-channel, are shown by the examples in Fig. 8.8(d) to (f). Bipolar transistors have three leads called emitter (e), base (b) and collector (c); and the leads of an FET are source (s), gate(g) and drain (d). These transistors are housed in a metal or plastics package. In many of the 'metal can' types, e.g. the TO18 shape, the collector lead is also connected to the metal can. Transistors used for high power applications have a flat metal side to them, e.g. the TO220 shape, to which a 'heat sink' can be bolted to help the transistor dissipate excess heat produced within it. The symbols for bipolar and unipolar transistors are shown in Fig. 8.9. Two of these transistors are used more than the others, namely the npn bipolar transistor and the n-channel field-effect transistor, and it is these two that we shall look at more closely, first as electronic switches and then as audio amplifiers.

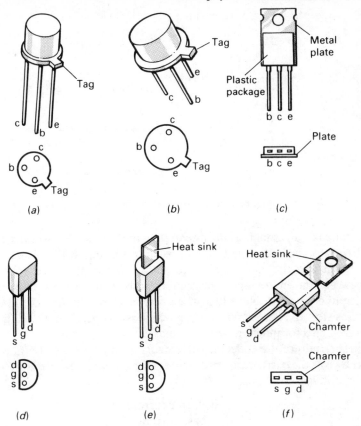

Fig. 8.8 A selection of bipolar and unipolar transistors: (a) npn type BC108; pnp type BC478; outline TO18, (b) npn type 2N3053; pnp type BFX88; outline TO39, (c) npn type TIP31A; pnp type TIP32A; outline TO220(A), (d) n-channel junction gate FET, type 2N3819; shape TO92(D), (e) n-channel metal-oxide field-effect transistor (MOSFET), type VN10LM; shape TO237, (f) n-channel MOSFET, type VN46AF; shape TO202(B)

8.6 How bipolar transistors work

Fig. 8.10a shows a simple model of an npn bipolar transistor. The base terminal is connected to a very thin layer of p-type semiconductor which is sandwiched between two layers of n-type semiconductor. This npn transistor comprises two p–n junctions, and it is useful to think of it as made up of two diodes connected together as shown in Fig. 8.10b. When an npn transistor is acting as

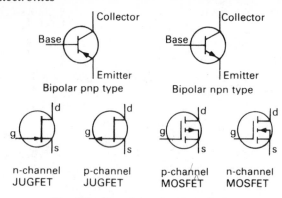

Fig. 8.9 Transistor circuit symbols

a switch or an amplifier, the direction of the (conventional) current flow into and out of its terminals is shown in Fig. 8.10c. A small current into the base terminal causes a much larger current flow between the emitter and collector terminals. Thus an npn transistor amplifies current. Note that when current flows through the npn transistor, the base–emitter p–n junction is forward-biased and the base–collector junction is reverse-biased.

The ratio of the collector current, I_c, to the base current, I_b, is known as the current gain of the transistor and is represented by the symbol, h_{FE}, so that

$$h_{FE} = \frac{\text{collector current}}{\text{base current}} = \frac{I_c}{I_b}$$

The current gain varies greatly from transistor to transistor, and even between transistors of the same type. It may be as little as 10

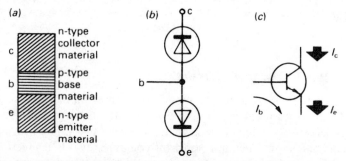

Fig. 8.10 An npn bipolar transistor: its (a) structure, (b) diode equivalent circuit, and (c) operating currents

for a high-power transistor such as one used in the output stages of a hi-fi amplifier, to as much as 1000 for a low-power transistor used in the input stages of a hearing aid. Fortunately, in most circuit designs it is possible to compensate for individual variations in current gain (see Section 8.9). A manufacturer merely guarantees that a particular device has a gain in a specified gain band, e.g. for a BC109 transistor the gain lies somewhere in the range 200 to 800. Note the relationship between the collector, emitter and base currents: $I_e = I_b + I_c$ since the current flowing into the transistor must equal the current flowing out of it. If the current gain is large, e.g. more than 100, I_b is much less than I_e or I_c so $I_c = I_e$, approximately.

The way an npn transistor amplifies current is quite complicated and rather difficult to understand, but a simple model should help you to grasp the general principles. Fig. 8.11a shows the distribution of electrons and holes in an npn transistor when the base terminal is left unconnected. The external power supply makes the collector terminal positive with respect to the emitter terminal. This external power supply reverse-biases the p–n junction formed by the collector and base regions of the transistor. Thus electrons and holes move away from the collector–base junction and no current flows across it.

Fig. 8.11b shows what happens when the npn transistor is working as a current amplifier. An external power supply now forward-biases the base–emitter junction and electrons flow easily across this junction. A few of these electrons combine with holes in the p-type

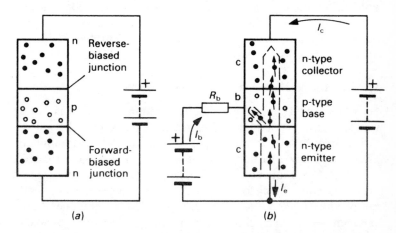

Fig. 8.11 Electrons and holes in a bipolar transistor: (a) transistor switched off, and (b) transistor switched on

base region. Not many electrons 'get lost' in this way, for two reasons: first, the base region is lightly doped with p-type impurities (Section 2.5) and, second, the base region is very thin. To make up for the 'loss' of electrons by recombination in the base region, an equal number of electrons leave the base terminal and these comprise the base current, I_b. Because the base region is very thin, most of the electrons which rapidly moved across the emitter–base junction are swept over the reverse-biased collector–base junction and form the collector current, I_c. In effect, the transistor has amplified the small base current to produce a larger collector current. Thus if 0.2% of the electrons that moved across the emitter–base junction combine with holes in the base, and the emitter current is 100 mA, the base current is

$$(0.2/100) \times 100 \text{ mA} = 0.2 \text{ mA}$$

Thus the current gain of the transistor is

$$\begin{aligned} I_c/I_b &= (I_e - I_b)/I_b \\ &= (100 - 0.2) \text{ mA}/0.2 \text{ mA} \\ &\simeq 500 \end{aligned}$$

Note that an npn transistor amplifies only when the base–emitter junction is forward-biased, i.e. when the base terminal is more positive than the emitter terminal by about 0.6 V (see Section 7.2). The reason why the npn transistor is called a 'bipolar transistor' is that both electrons and holes are responsible for its operation. Also note that practical transistor circuits show only conventional current flow into and out of the transistor. Thus the arrow inside the npn transistor symbol shows the conventional current flow across the forward-biased base–emitter junction of the transistor.

8.7 How the field-effect transistor (FET) works

As a discrete device, the field-effect transistor (FET) is not so widely used as the bipolar transistor though it has some characteristics that make it a better choice in electronic switches and amplifiers. These are its lower power consumption, its higher input resistance, and its higher frequency response. The lower power consumption makes it a good choice for integrated circuits, since excessive heat is not produced when many thousands of FETs are integrated on a small area of a silicon chip (Chapter 13). It is therefore possible to design portable battery-operated devices, e.g. personal computers and solar-powered calculators, using this type of integrated circuit.

Furthermore, FETs take up less space than bipolar transistors so more of them can be packed together on a silicon chip. In discrete form FETs look much the same as bipolar transistors (Fig. 8.8), but they work quite differently. There are two main types of FET, the junction-gate FET (JUGFET) and the more recent metal-oxide FET (MOSFET). There are 'n-channel' and 'p-channel' versions of each type. It is the n-channel version of these FETs which is described here, since it is more widely used than its p-channel counterpart. Let's look at the basic structure of each type of FET.

Fig. 8.12 shows the basic structure and circuit symbol of the JUGFET. It comprises a channel of n-type semiconductor through which electrons flow between the source terminal, s, and the drain terminal, d. This current, I_{ds}, is controlled by the gate-to-source voltage, V_{gs}. Note that the gate terminal is connected to a p-type semiconductor which is embedded in the n-type channel. Since the gate terminal is more negative than the source terminal, the p–n junction between the p-type gate and the n-type channel is reverse-biased and this causes a depletion region to extend into the n-type channel. Since the depletion region contains a negligible number of electrons (Section 7.2), electrons can only flow in what is left of the n-type channel. The resistance of this channel, and hence the size of I_{ds}, is determined by the width of the channel. And this width depends on the gate-to-source voltage, V_{gs}.

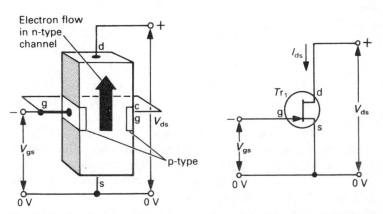

Fig. 8.12 Current flow through an n-channel junction-gate field-effect transistor (JUGFET)

Fig. 8.13 compares the width of the n-type channel for two different values of V_{gs}. The higher V_{gs}, the narrower the channel, the larger the channel resistance and the smaller I_{ds}. As V_{gs}

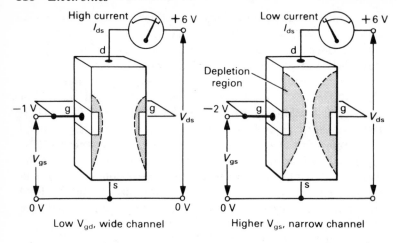

Fig. 8.13 How the gate-to-source voltage controls the channel width in a JUGFET

becomes more negative, the resistance of the channel increases. When V_{gs} has reached what is called the 'pinch-off voltage', the depletion region completely closes the channel and the resistance of the channel is very high. The pear-shaped structure of the depletion region shows that the depletion region is thickest at the drain end of the channel since here the p.d. between the drain and gate is greater.

Note the main difference between the operation of the npn bipolar transistor and the n-channel JUGFET: the bipolar transistor depends on the flow of both electrons and holes (hence 'bipolar'), but the n-channel JUGFET relies on the flow of electrons only, i.e. it is a unipolar device. (The p-channel FET relies on the flow of holes only.) Furthermore, whereas it is essential that a small flow of electrons leaves the base terminal of an npn transistor, a negligible and non-essential current leaves the gate terminal of an n-channel JUGFET. Thus the FET controls the flow of electrons through the n-channel by the electric field established between the gate terminal and the channel, i.e. by V_{gs}. That is why the 'field-effect transistor' is so-called. As a consequence, it has an input resistance which is considerably higher than that of a bipolar transistor.

Now compare the structure of the n-channel JUGFET with that of the n-channel MOSFET shown in Fig. 8.14a. In this structure, the gate terminal is insulated from the n-type channel by a layer of electrically-insulating silicon dioxide. This means that almost no current, much less than for a JUGFET, can flow to or from the gate terminal. There are actually two types of n-channel MOSFET, the depletion MOSFET and the enhancement MOSFET. In the latter device, a positive value of V_{gs} actually produces (i.e. enhances) an electron-rich n-channel just below the oxide layer. Thus the n-channel enhancement MOSFET is normally off until it receives a positive V_{gs}. Changes to V_{gs} cause changes to the electron density in this channel and hence changes in I_{ds}. Thus its operation is basically the same as an n-channel JUGFET except that the MOSFET requires a positive V_{gs} and the JUGFET a negative V_{gs}. The symbols for the MOSFETs shown in Fig. 8.8 are for the enhancement types. These n-channel and p-channel MOSFETS are also called n-MOS and p-MOS FETs.

One type of MOSFET which has become very popular in circuit design in recent years is the VMOS (vertical metal-oxide semiconductor) transistor. Though it has similar characteristics to the conventional MOSFET, its structure is quite different. Whereas an n-channel MOSFET has the drain, source and gate connections on the top of the silicon chip, Fig. 8.14b shows that a VMOS transistor

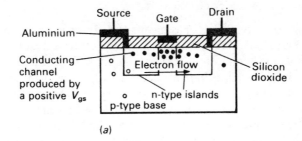

(a)

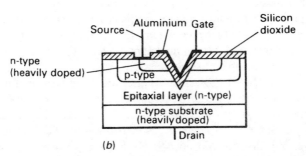

(b)

Fig. 8.14 The structure of two n-channel enhancement MOSFETs: (a) the basic MOSFET; (b) the VMOS transistor

has its source on the top of the chip and its drain at the bottom. The drain terminal is connected to a heavily doped n-type substrate on which a lightly doped n-type epitaxial layer is produced. Into this epitaxial layer is diffused first a lightly doped p-type region followed by a heavily doped n-type region for the source connection. Now a novel feature of a VMOS transistor is the V-groove which is etched through the p-type and n-type regions into the epitaxial layer. VMOS is an enhancement MOSFET so that current flows through it vertically from the drain to the source terminals along both sides of the V-groove, but only when the drain and gate terminals are positive with respect to the source terminal. The great advantage of this construction is that it provides a type of MOSFET which has higher voltage, current and power ratings than the conventional MOSFET. In addition, since the drain terminal is connected to the heavily doped n-type substrate, heat can be removed efficiently from the device via a heatsink bolted to the drain terminal. The circuit symbol for a n-channel VMOS device is the same as for an n-channel MOSFET. VMOS versions of the MOSFET are used in the circuit examples that follow.

8.8 Transistors as electronic switches

Despite the advantages of using integrated circuits in circuit design, millions of individual transistors are made each year, some of which are used in simple control circuits like the one shown in Fig. 8.15. This circuit switches on the lamp L_1 when the light-dependent resistor LDR_1 is covered, i.e. it is a dark-operated switch. Similar circuits are used in automatic street lights which come on at dusk. This electronic switch comprises three building blocks: a sensor based on R_1 and LDR_1 which make up a potential divider; a current amplifier based on transistor Tr_1; and an output device, in this case the lamp L_1. The on/off action of this circuit is controlled by the voltage at point X which rises when LDR_1 is covered, and falls when LDR_1 is illuminated. When the voltage at X rises, the lamp switches on; when it falls the lamp switches off.

Fig. 8.16a shows the action of the npn transistor as a current amplifier; small changes of base current produce large changes in the collector current. The ratio I_c/I_b is the current gain, i.e. 250 in this case. But I_b, and hence I_c, is controlled by the base–emitter voltage, V_{be}, as shown in Fig. 8.16b, a graph known as a **transfer characteristic**. If V_{be} is less than 0.6 V, I_b and I_c are both zero and the lamp is off. In this state, the collector–emitter resistance is very

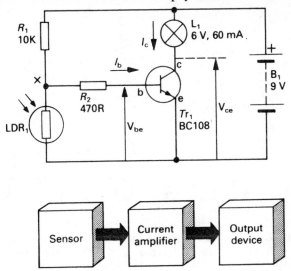

Fig. 8.15 A single transistor switching circuit comprising three building blocks

high and the collector–emitter voltage V_{ce}, is +9 V. The transistor is said to be 'cut-off' and the transistor switch is 'open'. As V_{be} increases above 0.6 V, I_b increases and the amplifying action of the transistor produces a much larger collector current, I_c. The lamp brightens until V_{be} has reached about 0.75 V. In this state, the collector–emitter resistance is so low that V_{ce} is nearly zero. The transistor is then saturated since any further increase in V_{be} does not increase I_c. The transistor switch is 'closed'.

One of the drawbacks with this simple one-transistor switch is that there is a small range of illumination of LDR_1 when the lamp is not fully switched on. However, if two transistors are used as shown in Fig. 8.17, the 'sharpness' of the switching action is improved considerably. This circuit is known as a **Darlington pair**. In this design, the emitter current of Tr_1 provides the base current of Tr_2. Also note that the voltage on the base of the first transistor must now exceed 2×0.6 V = 1.2 V before the lamp lights. The overall current gain of this combination of two transistors is the product of their individual current gains. Thus if the gain of each transistor is 100, the overall current gain is $100 \times 100 = 10\,000$. At least, this is the theoretical value, but for practical reasons the overall gain may be somewhat less than this. The collector current of Tr_1 is a hundred times smaller than the collector current of Tr_2 so that most of the

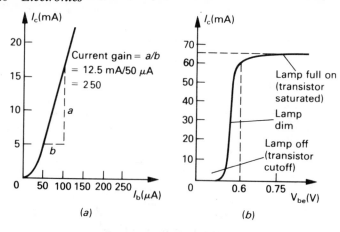

Fig. 8.16 Characteristics of an npn transistor: (a) variation of collector current with base current; (b) variation of collector current with base voltage

current flowing through L_1 flows through Tr_2. The overall effect is that the base current of Tr_1 is a hundred times smaller than when one transistor is used. This makes the circuit much more sensitive than the single transistor circuit, and lamps or relays are switched on and off smartly for small changes of light intensity.

Darlington pair transistors are available in discrete packages having just three leads: the base of the first transistor, the common

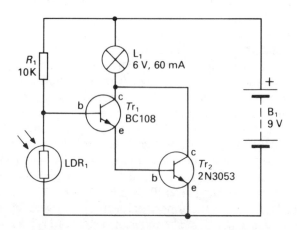

Fig. 8.17 A Darlington pair transistor switch

collector lead and the emitter of the second transistor. Indeed, they are also available in integrated circuit packages as shown in Fig. 8.18. This arrangement is ideal when microcomputers are used to control several motors or lamps via relays. The Darlington pairs amplify the low current signals available from a microcomputer's output lines.

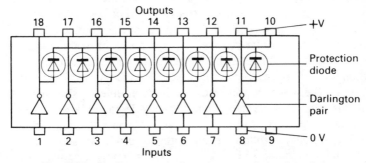

Fig. 8.18 Darlington pairs in an integrated circuit package

Fig. 8.19 shows that other sensors might be used in a Darlington-pair switching circuit. The variable resistor VR_1 enables some of the circuits to be set to switch at a predetermined level. Thus a 'rain alarm' is obtained using sensor S_1 which comprises a grid of conductors separated by insulation. If moisture bridges the gaps between the conductors, the relay energises. The use of a thermistor, Th_1 (Chapter 5), turns the circuit into a thermostat which could be used to maintain the temperature of a tropical fish tank at, say,

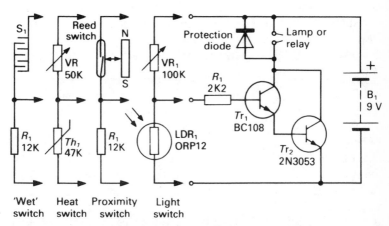

Fig. 8.19 Options for using a Darlington pair

30 °C. A proximity switch is obtained using a magnetically-operated reed switch (Chapter 4). Note that if any one of these sensors and the resistor in series with it change places, the circuits have the opposite switching function. Thus if LDR_1 is interchanged with VR_1, the circuit becomes a light-operated switch and switches on the relay when LDR_1 is illuminated.

Incidentally, note that in transistor switching circuits a silicon diode is connected across the relay. The diode is reverse-biased relative to the power supply, i.e. its anode is connected to the collector terminal. Its purpose is to protect the transistor from possible damage due to the high 'back e.m.f.' generated by the relay when it de-energises. Since the diode is reverse-biased it causes no drain on the power supply. But the diode harmlessly short-circuits the back e.m.f. generated when the relay de-energises since this is directed in the forward-bias direction, i.e. from collector to power supply, anode to cathode. These protection diodes are shown in Fig. 8.18 and are connected between the power supply terminal (pin 10 of the IC) and each of the collector terminals (pins 11 to 18) of the Darlington pairs in the IC package.

Though an improvement on the single transistor, the Darlington pair design is still not an ideal switching circuit. There is still a range of illumination, or of temperature, or of 'wetness', when the lamp is 'half on and half off'. This may lead to relay 'chatter' in which the contacts of the relay open and close rapidly just when they are supposed to snap open or closed. The problem is avoided in the two-transistor circuit shown in Fig. 8.20a, which is known as a **Schmitt trigger**. It has a snap-action response, the lamp (or relay) switches on smartly at one value of the input signal and switches off at a slightly lower value of input signal. Note that the voltage on the base of Tr_1 is controlled by the potential divider action of the three resistors R_1 to R_3. The operation of the Schmitt trigger is as follows:

(*a*) Assume, first, that LDR_1 is illuminated so that Tr_1 is switched off. This makes the collector voltage of Tr_1 high, which switches on Tr_2 and the lamp is lit. The emitter current of Tr_2 flows through R_4 which raises the voltage on the emitter of Tr_1 and reinforces its off state.

(*b*) If LDR_1 is slowly covered, Tr_1 begins to turn on. The collector voltage of Tr_1 begins to fall and with it the base voltage of Tr_2. The reduction in the current flowing through R_4 lowers the emitter voltage of Tr_1 which encourages Tr_1 to switch on. In turn, this helps to reduce the collector voltage of Tr_1 and hence switch Tr_2 further off.

(*c*) So the switching on of Tr$_1$ is helped by the switching off of Tr$_2$. The effect escalates and Tr$_2$ turns off smartly and Tr$_1$ turns on smartly, each transistor positively helping the other. The effect is called **positive feedback** and applies equally well when LDR$_1$ is subsequently slowly uncovered.

(*d*) Thus as more light reaches LDR$_1$, Tr$_1$ begins to switch off. Its collector voltage rises and tends to turn on Tr$_2$. The increase in the Tr$_2$ emitter current flowing through R_4 tends to raise the emitter voltage of Tr$_1$, helping it to turn off. The positive feedback action forces Tr$_1$ off and Tr$_2$ on.

Fig. 8.20b shows graphically how the input voltage V_{in}, i.e. the voltage at the base of Tr$_1$, determines the output voltage V_{out}, i.e. the voltage at the collector of Tr$_2$. In region A, LDR$_1$ is initially illuminated, V_{in} is LOW, V_{out} is LOW, and L$_1$ is on. As the illumination of LDR$_1$ decreases, the input voltage rises. When it reaches a certain upper threshold voltage, V_2, positive feedback rapidly causes Tr$_2$ to switch off, Tr$_1$ to switch on and L$_1$ goes out. In region B, the input voltage can go higher and lower than V_2, without changing the state of the Schmitt trigger provided it doesn't go below a certain lower threshold voltage, V_1. If it does, positive feedback rapidly switches Tr$_2$ on and Tr$_1$ off and L$_1$ switches on again. In region C, L$_1$ remains on if V_{in} goes lower or higher than V_1 provided V_{in} does not rise above the upper threshold voltage. If it does, Tr$_2$ switches off, Tr$_1$ switches on and L$_1$ goes off again and the voltages are in region D. Thus the Schmitt trigger has two input

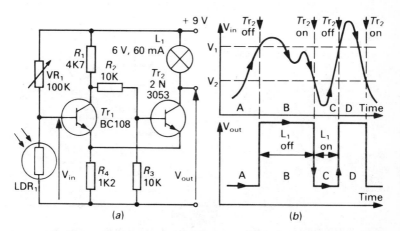

(a)   (b)

Fig. 8.20 (a) Making a Schmitt trigger from two transistors; (b) the switching action of the Schmitt trigger

voltages, the upper and lower threshold voltages between which there is no switching action. The difference between these two voltages, $V_2 - V_1$, is known as the *hysteresis* of the Schmitt trigger. Hysteresis produces a snap-action electronic switch just like the action of a mechanical wall light switch which has one position for on and one for off. A design for a Schmitt trigger based on an integrated circuit is described in Chapter 14.

Fig. 8.21a shows the general shape of the transfer characteristic of a VMOS device such as the VN10LM or VN46AF (see Fig. 8.8). This graph is quite different from the transfer characteristic (Fig. 8.16b) of a bipolar transistor. The npn bipolar begins to conduct when its base–emitter voltage is about 0.6 V, and an increase of about 0.2 V is sufficient to saturate the transistor. But the gate-to-source voltage, V_{gs}, required to make a VMOS device begin to conduct is much more variable, usually being between about 0.8 V and 2 V (known as the 'threshold voltage'). What is more, the value of V_{gs} needed to bring the device into saturation is very much higher than this, generally being in the region of about 10 V. Thus the drain current continues to increase over a wide range of gate voltage. In this respect, the VMOS device is inferior to the bipolar transistor as a switch. However, it does have one great advantage: whereas a high-power bipolar transistor might require a base current of, say, 100 mA to bring it into saturation, a VMOS device requires no current at all! It is voltage-operated, not current-operated.

A simple one-transistor switching circuit based on a VMOS transistor is shown in Fig. 8.21b. The lamp L_1 lights when the metal

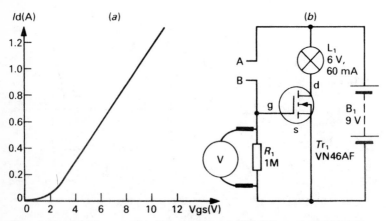

Fig. 8.21 (a) The transfer characteristic of a VMOS type of MOSFET;
(b) a VMOS 'touch switch'

contacts AB are bridged by a finger. The small current that flows through the finger raises the gate-to-source voltage V_{gs} above the threshold voltage. This voltage lowers the resistance of the channel so that I_{ds} increases and the lamp lights. If you connect a voltmeter as shown, you would find that the gate-to-source voltage, V_{gs}, rises and falls freely since no current flows into the gate terminal, i.e. the transistor has an extremely high input resistance. This is quite unlike the behaviour of a bipolar transistor: it switches on at 0.6 V and the base current that flows stops the voltage rising above 0.75 V. The high input resistance of a VMOS transistor can be inferred from the following experiment: remove resistor R_1 and comb your hair! The electrostatically-charged comb produces an electric field which induces a p.d. across the gate-to-source terminals of the VMOS transistor and changes the current flowing through the transistor without any current flowing into the gate terminal. The lamp varies in brightness as you move the comb near the circuit.

8.9 Transistors as audio amplifiers

Transistors are central to the design of audio amplifiers, whether they are used individually as discrete components, or as part of an integrated circuit on a silicon chip (see Section 8.2). Fig. 8.22a illustrates the design of a simple one-transistor voltage amplifier. It is suitable for 'listening in' to the 'sounds of light'. That is, it will detect the modulated light emitted from an a.c. mains-operated

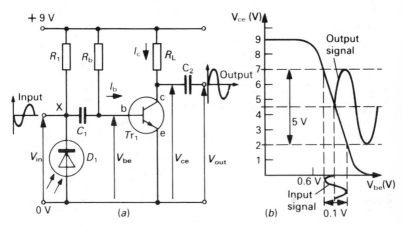

Fig. 8.22 A one-transistor voltage amplifier: (a) the circuit; (b) a graphical way of showing how voltage amplification occurs

lamp – and produce a 100 Hz buzz! More usefully, this voltage amplifier will detect the modulated light carrying information from a light transmitter in a simple optical communications system.

The photodiode D_1 detects the modulated light. This diode and the series resistor R_1 together act as a potential divider. Changes of light intensity cause small variations in voltage at point X. These changes are passed to the base of the transistor via the coupling capacitor C_1. The base resistor, R_b, ensures that a small steady (or 'quiescent') base current, I_b, flows across the base–emitter junction of the transistor. Now the variations in the base–emitter voltage V_{be} cause small variations in I_b. The transistor amplifies this current to produce larger variations in I_c. But the collector current flows through the 'load resistor', R_L, and provides an amplified output voltage V_{ce}, which is available via the second coupling capacitor, C_2. (By 'load resistor' is meant the resistor through which the collector current flows to produce the output voltage; without this 'load' the transistor would not act as a voltage amplifier.)

Fig. 8.22b shows graphically how this amplifier works. Small changes to the base–emitter voltage V_{be} (the input voltage) produce larger changes to the collector–emitter voltage V_{ce} (the output voltage). Thus the voltage gain of this amplifier is given by

$$\frac{\text{change in } V_{ce}}{\text{change in } V_{be}} = \frac{5 \text{ V}}{0.1 \text{ V}} = 50$$

Note that the d.c. (or quiescent) collector voltage, V_{ce}, of the transistor is set at half the supply voltage, i.e. 4.5 V in this case. This ensures that the changes to V_{ce} can 'swing' equally 2.5 V above and below the quiescent value of V_{ce}. If V_{ce} is set to a higher or lower value, the upper or lower part of the signal could be 'clipped'. This means the signal would be distorted, i.e. not be an amplified copy of the input signal. For example, if the steady collector voltage V_{ce} is set at 7 V, the same input signal would cause the upper part of the output signal to 'clip' as it would try to reach 7 V + 2.5 V, i.e. 9.5 V which is more than the e.m.f. of the power supply. Similarly, if V_{ce} is set at 2 V, the lower part of the output signal would be 'clipped'.

The setting of V_{ce} at half the supply voltage is achieved by choosing an appropriate value for the base resistor R_b. This resistor determines the base current, I_b, and therefore the value of collector current, I_c. Working backwards, let's first decide the value of the steady collector current I_c for the particular transistor we are using. For example, if it is a BC108, a value of 3 mA might be chosen. If the BC108 has a current gain h_{FE} of 150, the value of I_b is given by

$$I_b = \frac{I_c}{h_{FE}} = \frac{3\,\text{mA}}{150}$$

$$= 0.02\,\text{mA}\ (20\,\mu\text{A})$$

Now the value of R_b can be calculated. Since the supply voltage is 9 V, and the base–emitter voltage of this silicon transistor is 0.6 V, the p.d. across R_b is 9 V − 0.6 V = 8.4 V. Therefore

$$R_b = \frac{8.4\,\text{V}}{I_b} = \frac{8.4\,\text{V}}{0.02\,\text{mA}} = 420\,\text{k}\Omega$$

Thus a preferred value of 430 kΩ or 390 kΩ would be suitable for R_b.

Now the value of resistor R_L can be calculated. Since the d.c. output voltage has to be set at 4.5 V, and the collector current is 3 mA,

$$R_L = \frac{4.5\,\text{V}}{3\,\text{mA}} = 1.5\,\text{k}\Omega$$

This completes the design for this simple one-transistor voltage amplifier, but there are a couple of reasons why it is not a good design. First, it does not take into account the fact that the current gain of transistors, even if they are of the same type, varies greatly. Thus a small change in current gain for a given value of R_b would change I_b and hence I_c. For example, if h_{FE} increases to 200, the value of I_c increases to $I_b \times h_{FE} = 0.02\,\text{mA} \times 200 = 4\,\text{mA}$. Therefore the quiescent value of V_{ce} decreases to

$$9\,\text{V} - I_c \times R_L = 9\,\text{V} - 4\,\text{mA} \times 1.5\,\text{k}\Omega = 3\,\text{V},$$

i.e. V_{ce} falls from 4.5 V to 3 V and is not independent of transistor gain. And, second, if the temperature of the transistor rises, it produces an increase in the number of thermally-generated electrons and holes (Chapter 2) which increases the collector current. This current raises the temperature of the transistor and the effect escalates. This fault is called thermal runaway and is less of a problem with silicon transistors than with the older germanium transistors.

Fig. 8.23 is an improved design of voltage amplifier that stabilises the d.c. operating point of the transistor and increases its tolerance to variations in transistor gain. Instead of R_b being connected directly to the power supply line, its value is halved and it is connected to the collector terminal of the transistor. This change makes the quiescent base current I_b dependent on the quiescent

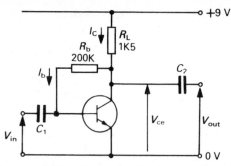

Fig. 8.23 Using negative feedback to make the voltage amplifier tolerant of changes in gain

collector voltage. The effect of this alteration to the design when the transistor's gain changes is as follows. Suppose the gain of the transistor increases. This increase tends to increase I_c but, as I_c increases, the value of V_{ce} falls due to the increase in p.d. across R_L. This reduces I_b, which tends to lower I_c and oppose the lowering of V_{ce}. Conversely, if the gain of the transistor falls, the effect is to reduce I_c, which raises V_{ce}. But this rise increases I_b and hence I_c which opposes the rise in V_{ce}. This technique of designing a circuit which compensates for variations in V_{ce} is one example of **negative feedback**, an important concept in electronics and one that is explained further in Chapter 14. The value of R_b in this design is calculated as follows.

$$V_{ce} = 9\,\text{V} - I_c \times R_L$$

But $I_c = h_{FE} \times I_b$, therefore

$$V_{ce} = 9\,\text{V} - h_{FE} \times I_b \times R_L$$

And since $I_b = V_{ce}/R_b$,

$$V_{ce} = \frac{9\,\text{V}}{1 + h_{FE} \times R_L/R_b}$$

Let's see the effect on V_{ce} of increasing the transistor's gain from 150 to 200. First, with $h_{FE} = 150$, $R_L = 1.5\,\text{k}\Omega$, and $R_b = 200\,\text{k}\Omega$,

$$V_{ce} = \frac{9\,\text{V}}{1 + 150 \times 1.5\,\text{k}\Omega/200\,\text{k}\Omega}$$

$$= \frac{9\,\text{V}}{2.125}$$

$$= 4.24\,\text{V}$$

Second, with $h_{FE} = 200$ and the same values of R_b and R_L,

$$V_{ce} = \frac{9\,\text{V}}{1 + 200 \times 1.5\,\text{k}\Omega/200\,\text{k}\Omega}$$

$$= 3.6\,\text{V}$$

Now this fall of 0.64 V in V_{ce}, from 4.24 V to 3.6 V, is not nearly as large as it is for the basic voltage amplifier above for which V_{ce} fell by 1.5 V from 4.5 V to 3 V. In fact, this stabilised circuit is a good general-purpose voltage amplifier, though even in this form it has one further drawback! This is that even when it is not amplifying a signal, it draws a current from the power supply (the quiescent current of 3 mA in the above example) and that's wasteful of electricity. However, we cannot avoid this since we had to set the collector voltage at half the supply voltage so that the load current could swing both up and down to cope with the rising and falling halves of the amplified signal. These amplifiers are called class A amplifiers.

A better design is one called a class B amplifier that only draws current when it is actually amplifying a signal. A rudimentary design of class B amplifier is shown in Fig. 8.24. It uses two 'complementary' transistors, an npn transistor, Tr_1, and a pnp, Tr_2, which have identical gains and power handling characteristics. The npn transistor handles the positive half-cycles of the input signal and the pnp transistor handles the negative half-cycles. This design is sometimes known as a 'push–pull amplifier' since Tr_2 'pushes' the negative half-cycles and Tr_1 'pulls' the positive half-cycles. Note

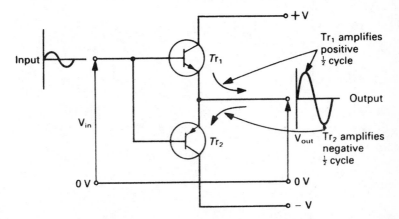

Fig. 8.24 The basic design of a push–pull amplifier

that the load, e.g. a loudspeaker, is connected in the common emitter circuit of the push–pull amplifier. When the input signal is positive, current flows down through this load from the emitter of Tr_1 to the 0 V power line; when the signal is negative, current flows up through this load from the 0 V power line to the emitter of Tr_2.

Of course, field-effect transistors can also be used in the design of audio amplifiers as shown in Fig. 8.25. In this design, an integrated circuit operational amplifier, IC_1, is used as a preamplifier followed by a VMOS transistor, Tr_1, used as a power amplifier. The function and use of operational amplifiers (op.amps) is described in Chapters 14 and 15. The quiescent output voltage of the op.amp. at pin 6 is set at half the supply voltage, i.e. 4.5 V, by the equal-value resistors R_1 and R_2 connected to its input pin 3. Thus the VMOS transistor Tr_1 is normally conducting, which makes this audio amplifier a class A design. An input signal applied to pin 3 of IC_1 via the coupling capacitor C_1 makes the gate voltage of Tr_1 rise and fall about 4.5 V. This allows Tr_1 to conduct more or less, so the source current through the loudspeaker, LS_1, rises and falls, producing an amplified audio output of the input signal. The variable resistor VR_1 acts as a volume control by setting the gain of the preamplifier based on IC_1.

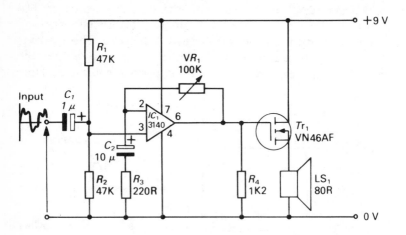

Fig. 8.25   A class A audio amplifier using a VMOS transistor

9

Logic Gates and Boolean Algebra

9.1 What logic gates do

The basic building blocks that make up all digital systems are simple little circuits called logic gates. A logic gate is a decision-making building block which has one output and two or more inputs as shown in Fig. 9.1. Two examples of these decision-making logic gates were introduced in Section 3.6. These were the AND gate and the OR gate using switches. The input and output signals of these gates can have either of two values, binary 1 or 0. The value of the output of a gate is decided by the values of its inputs. The truth table for a logic gate shows the value of the output for all possible values of the inputs. AND and OR gates (and the other gates described below) are known as logic gates because their outputs are the logical (i.e. predictable) result of a particular combination of input states.

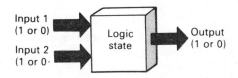

Fig. 9.1   A logic gate has two or more inputs and one output

Logic gates are used in computer, control and communications systems, and especially in calculators and digital watches. Fig. 9.2 shows the usual place of logic gates in digital systems. They have the intermediate task of receiving signals from sensors such as keyboard switches and temperature sensors, making decisions based on the information received, and sending an output signal to a circuit which provides some action, such as switching on a motor. Though

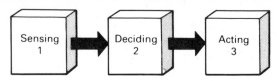

Fig. 9.2 The function of a digital system: logic gates are used in building block 2

decision-making building blocks in digital systems can be designed using individual diodes and transistors, or even mechanical switches as explained in Section 3.6, most digital circuits now make use of logic gates in integrated circuit packages. There are two main 'families' of these digital ICs; one is called transistor–transistor logic (TTL), and the other complementary metal-oxide semiconductor logic (CMOS). The operating characteristics of these two families are compared in Section 9.5.

9.2 Symbols and truth tables of logic gates

Two alternative systems are in use for showing the symbols of logic gates in circuit diagrams, the American 'Mil Spec' system and the British Standards system. Fig. 9.3 summarises these symbols for six logic gates, the AND, OR, NOT, NAND, NOR and Exclusive-OR gates. The American Mil Spec symbols are widely preferred, since their different shapes are easily recognised in complex circuit diagrams. But there is some pressure in the United Kingdom to adopt the British Standards symbols.

The AND gate gives an output of logic 1 when all inputs are at logic 1, and an output of logic 0 if any or all inputs are at logic 0. Thus an AND gate is sometimes called an 'all-or-nothing gate'. For the 2-input AND gate shown in Fig. 9.3a, the output, S, is at logic 1 only when input A and input B are at logic 1. The truth table for the 2-input AND gate gives the state of the output, S, for all combinations of input states – hence the term 'combinational logic' is used to describe logic systems using gates like the AND gate.

For the 2-input OR gate shown in Fig. 9.3b, the output, S, is at logic 1 when either input A or input B, or both inputs, are at logic 1. Thus the OR gate is sometimes called an 'any-or-all' gate. The truth table for the 2-input OR gate gives the state of the output, S, for all combinations of input states.

For the NOT gate shown in Fig. 9.3c, the output, S, is simply the

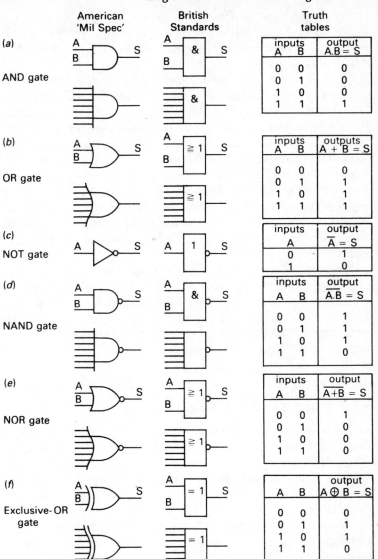

Fig. 9.3 Symbols and truth tables for logic gates

inverse of the input A. Thus if the state of the input is logic 1, the output state is logic 0 and vice versa. For this reason the NOT gate is also called an **inverter**. The truth table for the NOT gate is simple – it has only two lines.

The NAND (or NOT-AND) gate gives an output which is the converse of the AND gate. Thus for the 2-input NAND gate shown in Fig. 9.3d, the output, S, is at logic 1 when either input A or input B, or both inputs, are at logic 0. The truth table for the 2-input NAND gate gives the state of the output, S, for all combinations of input states. Compare it with the truth table of the AND gate.

The NOR (or NOT-OR) gate gives an output which is the converse of the OR gate. Thus for the 2-input NOR gate shown in Fig. 9.3e, the output, S, is at logic 0 when either input A or input B, or both inputs, are at logic 1. The truth table for the 2-input NOR gate gives the state of the output, S, for all combinations of input states.

The 2-input Exclusive-OR (or XOR) gate shown in Fig. 9.3f does something that the OR gate of Fig. 9.3b does not: it is a true OR gate for it only gives an output of logic 1 when either, but not both, of its inputs are at logic 1. The truth table summarises the action of the XOR gate which you should compare with that for the OR gate.

9.3 Introducing Boolean algebra

In 1847, George Boole invented a shorthand method of writing down combinations of logical statements which are either 'true' or 'false'. Boole proved that the binary or two-valued nature of his 'logic' is valid for symbols and letters as well as for words. His mathematical analysis of logic statements became known as Boolean algebra.

Now since digital circuits deal with signals which have two values, Boolean algebra is an ideal method for analysing and predicting the behaviour of these circuits. Thus 'true' can be regarded as logic 1, i.e. an 'on' signal, and 'false' as logic 0, i.e. an 'off' signal. But Boolean algebra did not begin to have an impact on digital circuits until 1938 when Shannon applied Boole's ideas to the design of telephone switching circuits.

The table below shows how Boolean algebra summarises the functions of the five logic gates, AND, OR, NOT, NAND, NOR and XOR, discussed above.

Logic statement	Boolean expression
(a) AND gate input A and input B = output S	$A.B = S$
(b) OR gate input A or input B = output S	$A + B = S$
(c) NOT gate not input A = output S	$\bar{A} = S$
(d) NAND gate not input A and input B = output S	$\overline{A.B} = S$
(e) NOR gate not input A or input B = output S	$\overline{A+B} = S$
(f) Exclusive-OR gate input A or input B = output S (excluding input A and input B)	$A \oplus B = S$

Note the use of the following conventions in Boolean algebra:

(a) two symbols are ANDed if there is a 'dot' between them;
(b) two symbols are ORed if there is a '+' between them;
(c) a bar across the top of a symbol means the value of the symbol is inverted.

The use of the bar across the top of a symbol is very important in Boolean algebra. Since the values of A and B can be either 0 or 1, $\bar{1} = 0$ and $\bar{0} = 1$ as in the NOT gate. If the bar is used twice, this represents a double inversion. Thus $\bar{\bar{1}} = 1$ and $\bar{\bar{0}} = 0$. What happens to the value of the symbol if the bar is used three times?

9.4 The universal NAND gate

It is possible to use one or more NAND gates to produce the logic functions of AND, OR, NOT, NOR and XOR. That is why a NAND gate is called a 'universal logic gate'. The more expensive NOR gate has the same property. Boolean algebra can be used to show that a NAND gate is 'universal' in this way.

To make a NOT gate, suppose we connect together the two inputs of a NAND gate as shown in Fig. 9.4a, and apply a logic 1 to these two inputs. Now for a NAND gate, we can write the Boolean expression $\overline{A.B} = S$. But A = B = 1, so $\overline{1.1} = \bar{1} = 0$. And if A = B = 0, $\overline{0.0} = \bar{0} = 1$. So a NOT can be obtained by connecting together the two inputs of a 2-input NAND gate and it then inverts the input signal. In fact, it doesn't matter how many inputs the NAND gate has: if all its inputs are connected together, the result is a NOT gate.

To make an AND gate, we need two NAND gates connected as shown in Fig. 9.4b where the NAND gate is followed by a NOT gate. The output of the NAND (not-AND) gate is followed by an inverter, a NOT gate, to produce an AND gate. Thus if R is the output from the first NAND gate, $R = \overline{A.B}$. Hence $S = \overline{R} = \overline{\overline{A.B}}$, or $A.B = S$. The two bars signify a double inversion which leaves the expression unchanged. The following truth table summarises the action of this combination of NAND gates.

A	B	$\overline{A.B} = R$	$S = \overline{R}$
0	0	1	0
0	1	1	0
1	0	1	0
1	1	0	1

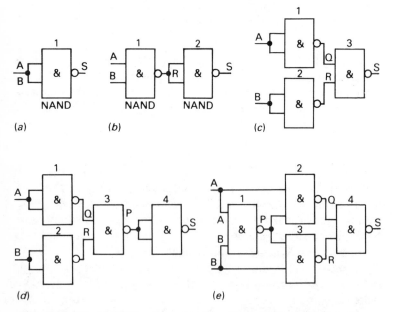

Fig. 9.4 How NAND gates are used to produce other logic functions: (a) NOT gate:$S=\overline{A}$ (b) AND gate:$S=A.B$ (c) OR gate:$S=A+B$ (d) NOR gate:$S=\overline{A+B}$ (e) XOR gate:$S=A\oplus B$

Note that the last column is the same as that of an AND gate (Fig. 9.3a).

To make an OR gate, three NAND gates are needed as shown in Fig. 9.4c. NAND gates 1 and 2 are each connected as NOT gates. Thus inputs Q and R to NAND gate 3 are: $Q = \overline{A}$ and $R = \overline{B}$. Thus $\overline{\overline{A}.\overline{B}} = S$. The truth table for this combination is as follows:

A	$\overline{A}$	B	$\overline{B}$	$\overline{A}.\overline{B}$	$\overline{\overline{A}.\overline{B}}$
0	1	0	1	1	0
0	1	1	0	0	1
1	0	0	1	0	1
1	0	1	0	0	1

Note that the last column is the same as that of an OR gate (Fig. 9.3b). Thus the two Boolean expressions, $\overline{\overline{A}.\overline{B}}$ and $A + B$ are equivalent to each other.

To make a NOR gate, the preceding OR gate must be followed by a NOT gate as shown in Fig. 9.4d. This means that a seventh column is added to the NAND gate truth table, and a fourth column is added to the truth table of the OR gate so that

$\overline{\overline{\overline{A}.\overline{B}}}$	$\overline{A+B}$
1	1
0	0
0	0
0	0

Since the last two columns are identical, we have to conclude that the combination of NAND gates in Fig. 9.4d performs the same function as a NOR gate. Moreover, the two Boolean expressions, $\overline{\overline{\overline{A}.\overline{B}}}$ and $\overline{A+B}$, are equivalent to each other.

It is more difficult to prove that the arrangement of NAND gates shown in Fig. 9.4e is equivalent to an Exclusive-OR gate. But if you work through the following truth table, you will find the solution very logical! The last two columns are identical, which proves that the combination of NAND gates in Fig. 9.4e is indeed equivalent to an XOR gate (Fig. 9.3f).

A	B	P = $\overline{A.B}$	$A.\overline{A.B}$	$B.\overline{A.B}$	Q = $\overline{A.\overline{A.B}}$	R = $\overline{B.\overline{A.B}}$	S = $\overline{(A.\overline{A.B}).(B.\overline{A.B})}$	
0	0	1	0	0	1	1	0	
0	1	1	0	1	1	0	1	(same as XOR
1	0	1	1	0	0	1	1	gate (Fig. 9.4e))
1	1	0	0	0	1	1	0	

9.5 Logic families

A logic 'family' is a particular design of logic integrated circuit such that all members of the family will happily work together in a circuit design. There are two main families of digital logic, the older of which is TTL (transistor–transistor logic) which was introduced by Texas Instruments Ltd in 1964. The TTL family has since been expanded to include many subfamilies, but the standard family is designated the '7400' series. For example, in this family the 7400 IC is a 'quad 2-input NAND gate', i.e. it contains four NAND gates each with two inputs and one output. Thus with two power supply pins, the 7400 IC has 14 pins ($4 \times 2 + 4 \times 1 + 2$) and is manufactured in the familiar dual-in-line (d.i.l.) package as shown in Fig. 9.5a. This device heads many other members of the 7400 family, some of which are listed in the table below.

Manufacturer's number	What it is
7400	quad 2-input NAND gate
7402	quad 2-input NOR gate
7404	hex inverter (i.e. six NOT gates)
7408	quad 2-input AND gate
7430	8-input AND gate
7432	quad 2-input OR gate
7486	quad Exclusive-OR gate

The standard TTL family has been the most popular family of logic ICs ever developed, though one of its subfamilies known as 'low-power Schottky' offers some improvements in performance. All members of the TTL family are based on bipolar transistors and operate from a $+5$ V (±0.25 V) power supply.

If the logic elements inside an IC are made from MOSFETs rather than bipolar transistors, we have the CMOS family of logic

ICs. CMOS stands for complementary metal-oxide semiconductor logic and refers to the fact that two types of MOSFET are used to make the logic elements, an n-channel MOSFET combined with a p-channel MOSFET (Section 8.7). The best known of the CMOS family is the 4000 series which was introduced in 1968. For example, the 4001 device is designated a 'quad 2-input NOR gate'. Like its counterpart the TTL 7400 device, the 4001 is housed in a 14-pin d.i.l. package as shown in Fig. 9.5b. But unlike the TTL family, the 4000 family is capable of operating from any supply voltage in the range 3 V to 15 V, and to 18 V if the device has a letter 'B' (for 'buffered') after its coding. The 4001 heads the 4000 series, some of which are listed in the table below.

Manufacturer's number	What it is
4001B	quad 2-input NOR gate
4011B	quad 2-input NAND gate
4069B	hex inverter (i.e. six NOT gates)
4071B	quad 2-input OR gate
4081B	quad 2-input AND gate
4030B	quad Exclusive-OR gate

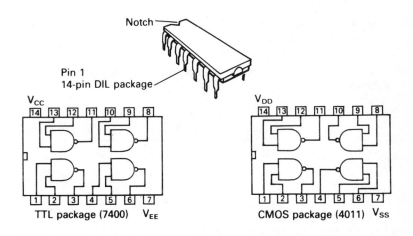

Fig. 9.5 TTL and CMOS versions of the quad 4-input NAND gate

One of the great advantages of CMOS is that CMOS devices require less electrical power than TTL devices. When a CMOS device is working, it dissipates about 1 mW per logic gate compared with about 15 mW for a standard TTL logic gate. Thus CMOS ICs are ideal for portable, battery-operated applications. Another advantage is that each MOSFET transistor that makes up CMOS logic circuits requires only about one fiftieth of the 'floor space' on a silicon chip compared with a bipolar transistor. So MOSFET ICs are ideal for complex silicon chips such as microprocessors and memories – see Chapter 13.

9.6 Decision-making logic circuits

The following three examples show how logic gates are used in decision-making circuits.

Example 1
Draw a logic circuit that satisfies the following conditions: a car can be started only if the gear lever is in neutral, the hand brake is on, and the seat belt is buckled.

First, let's decide that when each condition is satisfied, e.g. the belt is buckled, a logic 1 is input to the logic circuit. (The logic 1 signal could be the closure of, say, a reed switch.) The Boolean equation $S = A.B.C$ summarises the following function of the logic circuit:
car can be started (S) if
gear is in neutral (A)
AND hand brake is on (B)
AND belt is buckled (C)
The truth table for this Boolean equation contains eight lines as follows:

A	B	C	$S = A.B.C$	
0	0	0	0	
0	0	1	0	
0	1	0	0	
0	1	1	0	car cannot
1	0	0	0	be started
1	0	1	0	
1	1	0	0	
1	1	1	1	car can be started

A logic circuit that uses 2-input AND gates to satisfy this truth table is shown in Fig. 9.6. If universal 2-input NAND gates were available, each AND gate could be replaced by the logic circuit in Fig. 9.4b.

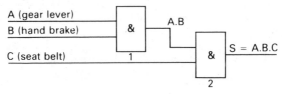

A (gear lever)
B (hand brake)
& A.B
C (seat belt)
1
& S = A.B.C
2

Fig. 9.6 Example 1

Example 2

An alarm system is to protect a room from unauthorised entry through a window, W, and a door, D. The alarm system is armed with a keyswitch, K, and the alarm sounds if either the keyswitch or the door is opened. Design a logic circuit that provides the protection required.

Let's assume that the keyswitch provides a logic 1 signal when the alarm system is armed, and the opening of the door or the window also provides a logic 1. The Boolean equation $S = K . (W + D)$ summarises the following function of the logic circuit:

the alarm sounds if the keyswitch is armed AND
the door is opened OR
the window is opened OR
the door and window are both opened.

The truth table for this Boolean equation is as follows:

K	W	D	$S = K . (W + D)$	
0	0	0	0	
0	0	1	0	
0	1	0	0	alarm does not sound
0	1	1	0	
1	0	0	0	
1	0	1	1	
1	1	0	1	alarm sounds
1	1	1	1	

A logic circuit that uses one 2-input OR gate and one 2-input AND gate is shown in Fig. 9.7. If only 2-input NAND gates were

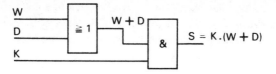

Fig. 9.7   Example 2

available, the circuit could be designed by using Fig. 9.4c to replace the OR gate and Fig. 9.4b to replace the AND gate.

Example 3

A chemical plant has four large tanks A, B, C and D that contain different liquids. Liquid level sensors are fitted to tanks A and B and temperature sensors to tanks C and D. Design a logic system that provides a warning when the level in tanks A or B is too high at the same time as the temperature in tanks C or D is too low.

First, let's decide what logic 1 and logic 0 signals represent:

Logic 1 is a signal that the level in tank A or B is too high, and logic 0 that the level is satisfactory.

Logic 1 is a signal that the temperature in tank C or D is too low, and logic 0 that the temperature is satisfactory.

A warning signal is represented by a logic 1 output from the circuit.

The Boolean equation $S = (A \oplus B) . (C \oplus D)$ summarises the following function of the logic circuit:

warning given if
level in tank A OR tank B is too high AND
temperature in tank C OR tank D is too low.

The truth table for this Boolean equation, which includes only those lines where $S = 1$ and a warning signal is produced, is as follows:

A	B	C	D	$A \oplus B$	$C \oplus D$	$S = (A \oplus B).(C \oplus D)$	
1	0	0	1	1	1	1	
1	0	1	0	1	1	1	warning is
0	1	0	1	1	1	1	given
0	1	1	0	1	1	1	

Note that we are using the Exclusive-OR condition, i.e. either A OR B and C OR D, and not A AND B and C AND D, must be at logic 1 to give an output of logic 1. For example:

A	B	C	D	$A \oplus B$	$C \oplus D$	$S = (A \oplus B).(C \oplus D)$	
1	1	0	0	0	0	0	warning not given
0	0	1	1	0	0	0	
1	1	1	1	0	0	0	

A logic circuit that uses two 2-input Exclusive-OR gates and one 2-input AND gate is shown in Fig. 9.8. If universal 2-input NAND gates were available, each Exclusive-OR gate could be replaced by four NAND gates as shown in Fig. 9.4e, and the AND gate by Fig. 9.4b. However, our decision to use Exclusive-OR gates ignores the possibility that the temperature in tanks A and B may be too low simultaneously, and/or the level in tanks C and D may be too high simultaneously. If these conditions are also to provide a warning, can you solve the above problem using NAND gates?

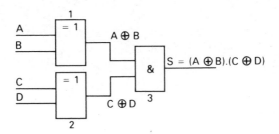

Fig. 9.8   Example 3

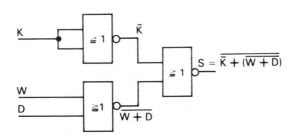

Fig. 9.9   An alternative solution for Example 2

Note that the aim of logic circuit design is to produce a circuit that uses as few universal logic gates as possible. Thus Example 2 could be solved using three 2-input NOR gates as shown in Fig. 9.9. Draw up a truth table to show that this logic circuit has the same function as the one given above.

10

Flip-Flops and Counters

10.1 What flip-flops do

You have seen that the truth table for a logic gate specifies the binary value (0 or 1) of the gate's output for every possible combination of binary input values. That is why a decision-making circuit using gates is generally called **combinational logic**. As well as being used in decision-making circuits, combinational logic circuits are the basic building blocks of encoders and decoders (Chapter 11), and of arithmetic circuits (Chapter 12).

Now a flip-flop is also made from logic gates, but its function is quite different from that of combinational logic circuits. A flip-flop 'remembers' its binary data until it is 'told' to forget it. The logic state of its output is determined by the value of the binary data it has stored in its memory, as well as any new data it is receiving. For this reason, circuits built from flip-flops are said to be **sequential logic** circuits. It is no surprise, then, that flip-flops are the basic memory cells in many types of computer memory (Chapter 13). They are also used in binary counters as explained in this chapter. The flip-flop is also known as a **bistable** and it is one of the family of multivibrator circuits which includes the monostable and the astable discussed in Chapter 6.

There are several different types of flip-flop, but the symbol for the most commonly used type is shown in Fig. 10.1a, a JK flip-flop. This type has two inputs labelled J and K (for no obvious reason). There is one input called the clock (CLK) input which is fed with on/off pulses, i.e. a series of 1s and 0s. (A clock in digital electronics is a pulse generator.) There are two outputs, Q and $\overline{Q}$, which are complementary, i.e. when output $\overline{Q}$ is at logic 1, output Q is at logic 0, and vice versa. In addition, the flip-flop has a SET input (S) and RESET input (R). A good example of a JK flip-flop in the '4000'

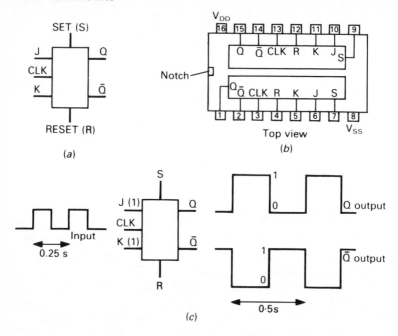

Fig. 10.1 The JK flip-flop: (a) its symbol; (b) the 4027 dual JK flip-flop; (c) waveforms produced when a flip-flop toggles

series of CMOS digital ICs is the 4027 device. As Fig. 10.1b shows, it contains two identical and independent flip-flops.

In this discussion we are interested in the use of the flip-flop as a **toggle flip-flop**. This means that when a series of clock pulses are fed into the clock input, each pulse causes the flip-flop outputs to change state from 1 to 0 or vice versa, i.e. to toggle. This behaviour is shown by the waveforms in Fig. 10.1c. To make the 4027 toggle, the J and K inputs must equal logic 1 and the S and R inputs must equal logic 0. Now suppose the frequency of the clock pulses is 4 Hz, i.e. the time between consecutive 1s or 0s is 0.25 s. This toggle flip-flop has the following three effects on the input waveform.

(*a*) The frequency of the waveform at each output has been halved by the flip-flop, i.e. the output frequency is now 2 Hz so that the time between consecutive 1s and 0s is 0.5 s.

(*b*) When the waveform at the Q output goes from HIGH to LOW, the waveform at the $\overline{Q}$ output goes from LOW to HIGH, and vice versa.

(*c*) The change in the state of the Q outputs takes place when the clock pulse changes from LOW to HIGH; we say the 4027 is positive-edge triggered.

The truth table below shows the HIGH (1) and LOW (0) states of the Q and $\overline{Q}$ outputs as the 4027 flip-flop is toggled by the clock pulses.

CLK	Q	$\overline{Q}$	
L(0)	H(1)	L(0)	
H(1)	L(0)	H(1)	a change
L(0)	L(0)	H(1)	no change
H(1)	H(1)	L(0)	a change
L(0)	H(1)	L(0)	no change
H(1)	L(0)	H(1)	a change

. . . . and so on

This table clearly shows that a change in the state of one of the two outputs occurs when the clock pulse changes from 0 to 1, and that this change occurs for every two HIGH-to-LOW changes of the clock pulse. Let's see how useful this behaviour of a toggle flip-flop is in the design of electronic counters.

10.2 Making binary counters from flip-flops

Suppose that every time a binary 1 enters a flip-flop it represents a single count. This count, for example, could be the signal produced by a reed switch every time a bicycle wheel rotates. Thus if sixteen counts are input to this flip-flop, eight counts are delivered from its Q output. Fig. 10.2 shows what happens if the Q output from this first flip-flop, FF$_1$, is connected to the clock (CLK) input of a second flip-flop, FF$_2$. Since each flip-flop divides by two, two flip-flops divide by four $(2 \times 2 = 2^2)$ and four counts leave FF$_2$. If a third flip-flop, FF$_3$, is connected to the Q output of FF$_2$, the overall division is eight $(2 \times 2 \times 2 = 2^3)$ and two counts leave the Q output of FF$_3$. A fourth flip-flop, FF$_4$, provides an overall division by sixteen (2^4) and one count leaves FF$_4$. Four flip-flops connected as shown in Fig. 10.2 are said to be **cascaded**. It is usual to label the four outputs Q$_A$ to Q$_D$. The pattern of 0s and 1s from these four outputs make up sixteen input counts as shown in the following table. The pattern of 0s and 1s from the Q outputs shows that eight counts leave

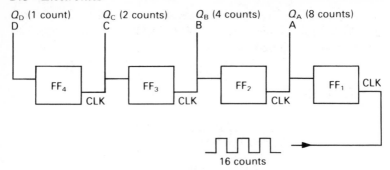

Fig. 10.2 A 4-bit binary counter using four cascaded flip-flops

Q_A for sixteen counts entering flip-flop, FF_1. Four counts leave Q_B, two counts leave Q_C, and one count leaves Q_D.

Input count	Q_D(MSB)	Q_C	Q_B	Q_A(LSB)
0	0	0	0	0
1	0	0	0	1
2	0	0	1	0
3	0	0	1	1
4	0	1	0	0
5	0	1	0	1
6	0	1	1	0
7	0	1	1	1
8	1	0	0	0
9	1	0	0	1
10	1	0	1	0
11	1	0	1	1
12	1	1	0	0
13	1	1	0	1
14	1	1	1	0
15	1	1	1	1
16 counts	1 count	2 counts	4 counts	8 counts

The four columns labelled Q_A to Q_D give the binary equivalent of the number of counts entering the first flip-flop, FF_1. Thus when eleven decimal counts have entered the first flip-flop, the binary value represented by the 1s and 0s of the flip-flop outputs is 1011. That this binary number is equivalent to decimal 11 can be found by adding the decimal values of the four binary digits, i.e.

$$1 \times 2^0 + 1 \times 2^1 + 0 \times 2^2 + 1 \times 2^3 = 11$$

Binary-to-decimal conversion is explained more fully in Chapter 12. Since four binary digits make up each number in the table, the counter is called a **four-bit binary counter**.

The binary digit (or **bit**, for the sake of brevity) in the Q_A column is known as the least significant bit (LSB) since it has the lowest binary weighting, i.e. 2^0, in the number. The bit in the Q_D column is known as the most significant bit (MSB) since it has the highest binary weighting, i.e. 2^4. Note that this binary counter counts to decimal 15, from binary 0000 to binary 1111. A five-bit binary counter able to count from 00000 to 11111 (decimal 31) is obtained by adding a fifth flip-flop after FF_4. Six flip-flops would count to binary 111111 (decimal 63), seven to binary 1111111 (decimal 127), and eight to binary 11111111 (decimal 255). Clearly, if a binary counter has n flip-flops, there are n bits in the number which can be counted.

A **binary-coded decimal (BCD) counter** is a version of the four-bit counter that counts to decimal 9 in binary. It is used in many types of digital display, e.g. in digital watches, weighing scales and petrol pumps. Three BCD counters can be connected as shown in Fig. 10.3 to display a binary count equivalent to the decimal number 139. Counter Z shows the binary value (1001) of the 'units' digit, counter Y the binary value of the 'tens' digit (0011), and counter X the value (0001) of the 'hundreds' digit. Note that the MSB of each counter is connected to the CLK input of the preceding counter. Every time the MSB changes from 1 to 0, a pulse is passed from counter Z to counter Y, or from counter Y to counter X. Initially all three BCD counters are set to 0000 by applying a reset pulse to all counters simultaneously. After 139 clock pulses, BCD counter Z shows a binary value of 1001, counter Y a binary value of 0011, and counter

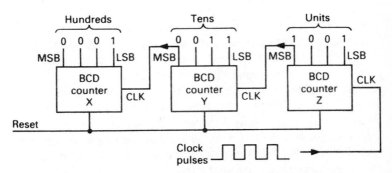

Fig. 10.3 Three cascaded BCD counters

X a binary value of 0001. How do these three four-bit numbers get there?

Suppose 9 pulses have entered counter Z. Counter Z then shows 1001 and counters Y and X each show 0000. After the next clock pulse, counter Z changes to 0000 and counter Y to 0001. For pulses 11 to 19, counter Y shows a steady reading of 0001 while counter Z passes through the binary counts of 0001 to 1001. After the 20th pulse, counter Y changes to 0010 and counter Z goes back to 0000. The same thing happens to counter Z after 29 pulses; counter Z changes from 1001 to 0000 and counter Y increases to 0011. Thus after 99 pulses, counters Z and Y show 1001 and counter X 0000. After the next clock pulse (decimal 100), counter Z and Y both show 0000 and counter X 0001. During the next 39 clock pulses counter X shows a steady 0001 while counters Z and Y repeat the changes for the first 39 pulses. The maximum number of clock pulses that can be displayed by these three BCD counters is 999. Chapter 11 explains how the binary numbers from BCD counters can be displayed in a decimal form on LED and LCD displays.

10.3 Frequency dividers and counters

Computers, digital clocks and watches, and many types of control system depend for their operation on a series of electronic 'ticks' which occur at a fixed frequency. For example, mains-operated digital clocks make use of the 50 Hz mains frequency which is maintained accurately at this value by the power stations. But microcomputers, digital watches and clocks use the rapid vibrations of a tiny quartz crystal (Fig. 10.4) to provide a stable frequency of, for example, 32.768 kHz. How can flip-flops and binary counters be used to reduce frequencies of 50 Hz and 32 kHz to a frequency of 1 Hz?

Fig. 10.5 shows the system used in a digital watch for reducing the 32.768 kHz generated by a quartz crystal to a frequency of 1 Hz. Here a set of cascaded flip-flops repeatedly divide the frequency by two. How many times should 32.768 kHz be divided by two? The answer is fifteen times since 2 multiplied by itself 15 times equals 32 768. These 1 Hz pulses are then delivered to a set of cascaded BCD counters followed by decoders (Chapter 11) so that a digital display can be operated. In a digital watch, the oscillator (except the crystal itself), the flip-flops, BCD counters and decoders are all contained in a single integrated circuit.

The reduction of the mains frequency to 1 Hz requires the system

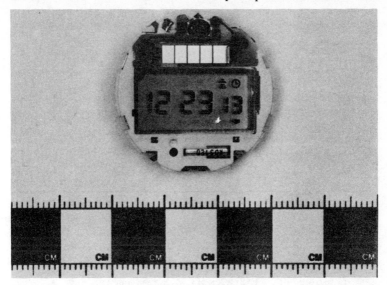

Fig. 10.4 The crystal in this watch is the bottle-shaped object below the liquid crystal display

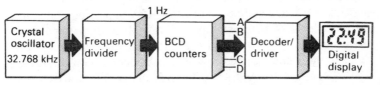

Fig. 10.5 The electronic system of a digital watch

shown in Fig. 10.6. First a BCD counter divides the mains frequency by 10 to give 5 Hz. Then a second BCD counter is arranged to divide by 5 to give the 1 Hz frequency required. These 1 Hz pulses are then used to operate BCD counters and decoders to drive the display.

The 4020, 4040 and 4060 devices shown in Fig. 10.7 are frequency dividers in the CMOS family of ICs. They contain several cascaded

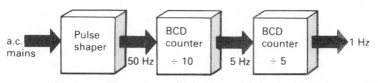

Fig. 10.6 The electronic system of a mains-operated clock

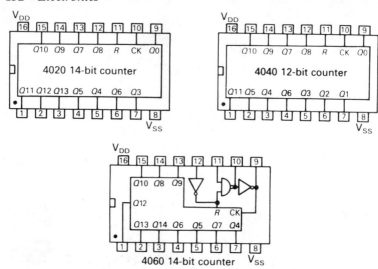

Fig. 10.7   Examples of CMOS frequency dividers

flip-flops to divide by 2^{14}, 2^{12} and 2^{14}, respectively. All the outputs of the internal flip-flops are available from the 4040 device so a maximum frequency division of $2^{12} = 4096$ is possible. Obviously the 4020 and 4060 devices cannot provide all flip-flop outputs since there aren't enough pins on the package. The 4020 misses out division by 2^2 and 2^8, and the 4060 misses out division by 2^1, 2^2, 2^3 and 2^{11}. The 4060 provides maximum frequency division by a factor of $2^{14} = 2 \times 2 \times 4096 = 16\,384$. Thus the 4040 could be used to divide the crystal frequency in Fig. 10.5 by 16 384 to provide an output frequency of $32\,768/16\,384 = 2$ Hz; one further flip-flop would provide 'ticks' at one second intervals.

These frequency dividers are easy to use. In normal use, the counter counts every time the input signal changes from HIGH to LOW. For counting to take place, the RESET input (e.g. pin 11 on the 4040) must be connected LOW, i.e. to 0 V. If the RESET is taken HIGH, the counters reset to zero, i.e. all outputs become logic 0.

The 4510 and 4516 CMOS devices shown in Fig. 10.8 are both 4-bit binary counters, the former a BCD counter with a maximum binary count of 1010 (decimal 9), and the latter a 4-bit counter that has a maximum count of 1111 (decimal 15). Both devices contain four cascaded flip-flops so they could replace the design based on individual flip-flops shown in Fig. 10.3. And both devices can count

Flip-Flops and Counters 153

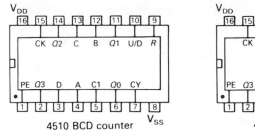

Fig. 10.8 Examples of CMOS 4-bit and BCD counters

(a)

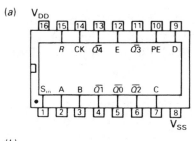

(b)

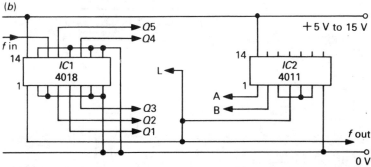

Divide by	Connect L to
2	Q1
4	Q2
6	Q3
8	Q4
10	Q5

Divide by	Connect A to	Connect B to
3	Q1	Q2
5	Q2	Q3
7	Q3	Q4
9	Q4	Q5

Fig. 10.9 The 4018 divide-by-*n* counter/divider: (a) its appearance, and (b) connections for frequency division

up or down in binary depending on whether the UP/DOWN terminal (pin 10) is set at logic 1 or logic 0, respectively. Individual four-bit counters can be cascaded by connecting the CARRY OUT (CO) pin of one device to the CARRY IN (CI) pin of the next device. When used as a binary counter, the CARRY IN, RESET and LOAD terminals are set at logic 0. There are other connections on these ICs which need not worry us here.

The 4020, 4040 and 4060 frequency dividers can only divide an input frequency by powers of 2, but the 4018 shown in Fig. 10.9 is a presettable 'divide-by-*n*' counter designed to divide by any whole number between 2 and 10. It contains five flip-flops and some complex logic circuitry. If the DATA terminal (pin 1) is connected to pins 3, 5, 7 and 9, a frequency input to pin 14 is divided by 2, 4, 6, 8 and 10 respectively. Division by 3, 5, 7 and 9 requires the use of a 4011 quad 2-input NAND gate.

A particularly useful counter is the CMOS 4017 device shown in Fig. 10.10. It is a decade counter, i.e. it has ten decimal outputs and an overflow, or CARRY OUT (CY), output. As the waveforms

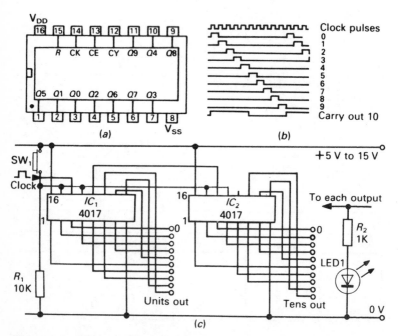

Fig. 10.10 The CMOS 4017 decade counter: (a) its appearance, (b) output waveforms, and (c) use as a two-digit counter

show, when its RESET terminal (pin 15) and CLOCK ENABLE (pin 13) terminal are at logic 0, these ten outputs go HIGH each time the clock pulse goes from LOW to HIGH. At any point in the counting sequence, all the LEDs except LED_0 (the 'zero' count on pin 3) can be switched off by momentarily connecting the RESET pin to logic 1.

The circuit in Fig. 10.10c shows how to wire up a simple decimal counter that counts to 99 on two banks of LEDs by cascading two 4017s. The CARRY OUT pulse from IC_1 (which measures the 'units' count) completes one cycle for every ten input pulses. This divide-by-ten pulse is used to operate IC_2 which measures the 'tens' count. Twenty LEDs with 1 kΩ series resistors show the count at the outputs.

11

Displays, Encoders and Decoders

11.1 Analogue and digital displays

You have only to think of the array of instruments in the cockpit of a
modern fighter, or the control room of a power station to realise that
the most convenient way of conveying information to a human
operator of an electronic system is to use some form of visual
display. The displays used are generally of two types, 'analogue' and
'digital'. These terms were used in Section 4.3 to describe analogue
and digital multimeters. An analogue multimeter displays the value
of a measurement on a moving-coil meter which uses a pointer that
moves over a calibrated scale. On the other hand, a digital multimeter generally uses a liquid crystal display (LCD) or a seven-
segment light-emitting diode (LED) display to give a numerical
value of a measurement.

Digital displays based on LCDs and LEDs have largely replaced
analogue displays in many different types of instrument. The main
advantage of LED and LCD displays is that, having no moving
parts, they are more rugged and can stand up to vibration better
than the rather fragile moving-coil meter. They are also cheaper and
easier to manufacture, and purpose-designed ICs are readily available to operate them. But perhaps the main reason for their rise to
fame is that many of today's electronic systems process signals which
are 'digital', and digital signals are compatible with the operating
principles of LCDs and LEDs. That is, LCDs and LEDs simply
need to be driven on and off by digital signals. That's why digital
displays are a natural choice for displaying information in all kinds
of electronic systems, from digital watches to microcomputers.

But remember, a numerical display is not always the preferred
choice in a digital system. Sometimes it is better to use an analogue
display when the change in a reading is looked for. Analogue

displays are often used on hi-fi amplifiers in preference to digital displays for indicating the audio power delivered to loudspeakers or the signal strength of a radio station. These analogue displays use a 'bar of light' made of discrete LEDs or LCD segments which lengthens or shortens in response to the signal strength. Analogue displays of this kind make it easier to see how the signal strength changes with time rather than giving a precise value. Perhaps that's why some people prefer digital watches with LCD 'hands' since the time of day seems to have more meaning when set against the twelve-hour time scale round the face of the watch. Fig. 11.1 shows how analogue and digital displays are combined in one digital instrument.

Fig. 11.1 Digital and analogue liquid crystal displays in one. The vertical dark line moves along the graduated scale
Courtesy: Megger Instruments Ltd

11.2 The seven-segment LED display

The combination of electronics and optics to display information is known as **optoelectronics** – an LED (Chapter 7) is an optoelectronic device. Seven LEDs are used in the seven-segment display shown in

Fig. 11.2a. Numbers, letters and other symbols are formed by the selective illumination of one or more segments arranged in the form of the figure '8'. Each of the LEDs labelled a to g can be switched on or off by digital circuits. A display of this type which forms both numbers and some letters is known as an alphanumeric display.

Fig. 11.2b shows how the decimal number 6 is produced by switching on segments c to g. The decimal number 1 is displayed by switching on segments b and c, and so on. Only the decimal numbers 0 to 9, a few special symbols such as '—', and a few letters such as C, c, and F can be displayed by these seven-segment displays. A decimal point can be displayed by illuminating an eighth LED to the right or left of the digit.

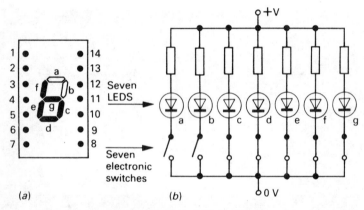

Fig. 11.2 A seven-segment display: (a) how the segments are labelled; (b) how decimal '6' is obtained

The digital circuits needed to light particular segments of the seven-segment display are described below. There are two types of seven-segment display depending on the nature of the digital circuits used. In the common-anode type shown in Fig. 11.3a, the anodes of all seven LEDs are connected to the positive terminal of the power supply. To light one of the segments, the cathode terminal of the segment LED is provided with a LOW signal by the digital circuits. In the common-cathode type shown in Fig. 11.3b, the cathodes of all seven LEDs are connected to the 0 V terminal of the power supply. To light one of the segments, the anode terminal of the segment LED is provided with a HIGH signal from the digital circuits.

As explained in Section 7.5, the LEDs are made from specially

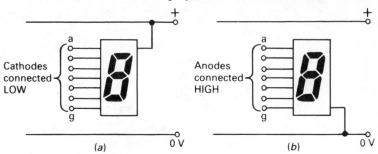

Fig. 11.3 (a) A common-anode seven-segment LED display requires a LOW drive signal, and (b) a common-cathode seven-segment LED display requires a HIGH drive signal

doped semiconductors, mainly gallium arsenide. The colour of the light emitted from the LED depends on the type of 'impurity' introduced into the crystal structure of gallium arsenide. In this way, LEDs that emit red, green, yellow and even blue light can be manufactured.

11.3 The liquid crystal display (LCD)

The LCD is a popular method of displaying information, especially in digital watches and pocket games. LCDs can display not just numerical data, but also words and pictures. Even some oscilloscopes and portable microcomputers now use large-area LCDs rather than a cathode-ray tube – see Fig. 11.4. The main reason for choosing LCDs for these applications is that their power consumption is minute compared with LED displays. Whereas the LED display requires electrical power to generate light, the LCD simply controls available light. This means that it is easily seen in bright sunlight but it cannot be seen in the dark. Hence watches and televisions like the one shown in Fig. 11.4 are provided with a 'backlight' for night-time viewing.

The LCD relies on the transmission or absorption of light by certain organic carbon crystals which behave as if they were both solid and liquid; that is, their molecules readily take up a pattern as in a crystal and yet flow as a liquid. In the construction of the common LCD unit shown in Fig. 11.5, this compound is sandwiched between two closely-spaced, transparent metal electrodes which are in the form of a pattern, e.g. as a seven-segment digit. When an a.c. signal is applied across a selected segment, the electric field

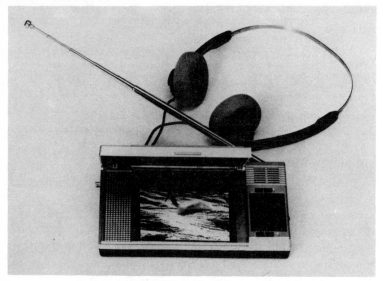

Fig. 11.4 An LCD screen in a miniature TV
Courtesy: Citizen Watch (UK) Ltd

causes the molecular arrangement of the crystal to change, and the segment shows up as a dark area against a silvery background. A polarising filter on the top and bottom of the display enhances the contrast of black against silver by reducing reflected light. This type of LCD is called a field-effect LCD, since it relies on the electric field produced by the a.c. signal. LCDs that produce frosty white

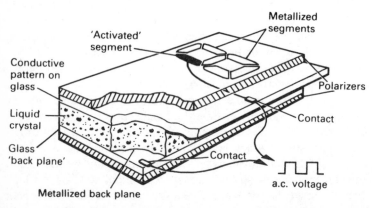

Fig. 11.5 The structure of a liquid crystal display

characters on a dark background are also available but are much less commonly used. This type of LCD is known as a dynamic scattering LCD. It uses a different liquid crystal and no polarisers, and it consumes more power than the field-effect LCD.

11.4 Decoders and encoders

In digital electronics, decoders and encoders are important output and input functions of a system. They are code translators, i.e. they change information from one form to another. Fig. 11.6 shows these two different functions:

a **decoder** is used to alter the format of information taken from a system, and

an **encoder** is used to alter the format of information entered into a system.

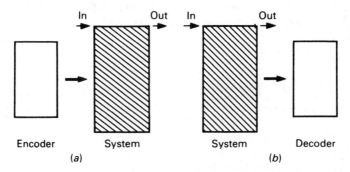

Fig. 11.6 In an electronic system (a) encoders are input devices and (b) decoders are output devices

An example of a decoder is a BCD-to-seven-segment decoder/driver. This device is required to change the 4-bit code produced by a BCD counter into signals capable of displaying decimal numbers on a seven-segment display. An example of an encoder is a telephone keypad which converts a telephone number into a string of binary digits for transmission along a telephone line. Decoders and encoders are examples of combinational logic devices, and it is quite possible to construct them from AND, XOR, etc. logic gates. However, purpose-designed integrated circuit packages are available to make the job easier for the circuit designer.

For example, decoders are used in seven-segment displays for converting 4-bit BCD codes into patterns of 1s and 0s for operating the LEDs in seven-segment LED displays. These decoders are known as seven-segment decoder/drivers. They are called decoder/drivers since they decode from binary to decimal and provide the necessary current to light, i.e. 'drive', the LEDs in the display.

As shown in Fig. 11.2, the seven segments of a seven-segment LED display are labelled a to g. In the range of TTL digital devices, the 7447 device is a decoder/driver which is designed to drive the segments on or off to display the decimal numbers 0 to 9. Fig. 11.7 shows the connections required between the 7447 decoder/driver and the segments of a common-anode display. The output terminals of the 7447 decoder/driver are connected to the corresponding cathode terminals of the segment LEDs. Note that in this common-anode display, all the anodes of the segment LEDs are connected together to +5 V. This means that the outputs a to g of the 7447 have to go LOW, i.e. to 0 V, to light the corresponding segments and produce a number. The outputs of the 7447 are said to be 'active LOW'. The 7447 'sinks' current (up to 40 mA per segment) into each of its output pins when that pin is LOW. Note that series

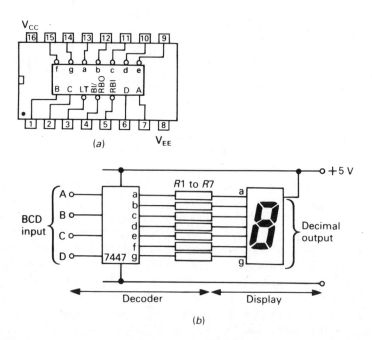

Fig. 11.7 The 7447 decoder/driver: (a) its pin connections, and (b) the connections to a common-anode seven-segment LED display

resistors are needed to limit the current flowing through the segment LEDs.

The CMOS equivalent of the 7447 decoder/driver is the 4511 device. The outputs of this decoder/driver are 'active HIGH' so it has to be used with a common-cathode display. Each output of the 4511 'sources' up to 5 mA of current when operated from a 10 V power supply. The 4511 decodes the ten 4-bit binary numbers, 0000 to 1001, and produces the following patterns of 0s and 1s for driving the seven segments of a common-cathode display. Note that the 4511 does not activate segment d when showing decimal 9, or segment a when showing decimal 6, as is normal in digital watches, clocks and calculators.

BCD inputs				Segment outputs							Number displayed
D	C	B	A	a	b	c	d	e	f	g	
0	0	0	0	1	1	1	1	1	1	0	0
0	0	0	1	0	1	1	0	0	0	0	1
0	0	1	0	1	1	0	1	1	0	1	2
0	0	1	1	1	1	1	1	0	0	1	3
0	1	0	0	0	1	1	0	0	1	1	4
0	1	0	1	1	0	1	1	0	1	1	5
0	1	1	0	0	0	1	1	1	1	1	6
0	1	1	1	1	1	1	0	0	0	0	7
1	0	0	0	1	1	1	1	1	1	1	8
1	0	0	1	1	1	1	0	0	1	1	9

In a liquid crystal display (LCD), there must be no d.c. signal across the display. So a special form of drive circuit is needed as shown in Fig. 11.8. The binary number 1001 (decimal 9) is being received by the BCD decoder (e.g. the CMOS 4543). The decoder therefore activates the c, d, e, f and g outputs – the decoder outputs are active HIGH. The a and b outputs are LOW. The backplane of the display receives a 30 Hz square wave signal which is also applied to each of the CMOS Exclusive-OR gates used to drive the LCD. The XOR gates ensure that when the segment input is HIGH, the segment drive voltage and the backplane voltage are exactly 180° out of phase, i.e. the inverse of each other. Thus there is an overall a.c. voltage across the liquid crystal, resulting in this segment being dark. When the segment input is LOW, the segment input and the backplane are exactly in phase. There is no longer a voltage across the liquid crystal and the segment remains transparent.

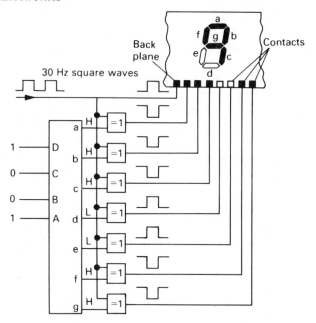

Fig. 11.8   The drive circuit required for an LCD

11.5 Multiplexing seven-segment displays

Each digit in a digital display requires one BCD counter and one decoder/driver as shown in Fig. 11.9 for a two-digit display based on TTL devices. In this circuit, when the 'units' digit changes from 9 to 0, the 'tens' digit increases by 1. In common with digital watches, clocks and other displays, this simple display has 'leading zero blanking' (LZB). This means that for counts below 10 the 'tens' digit is blanked, i.e. it does not show up as '0'. LZB saves on battery power and makes the display easier to read.

If we wanted to extend the simple two-digit display shown in Fig. 11.9 to three, four or more digits, the circuit becomes complicated since each digit requires its own decoder/driver. Also, all the digits are switched on at the same time, which is quite a drain on battery power. However, there is a technique which reduces circuit complexity by using just one decoder/driver, and which reduces the drain on the power supply by switching the digits on one at a time in rapid succession. The technique is called **multiplexing**.

Multiplexing scans the digits one after the other rapidly, so that

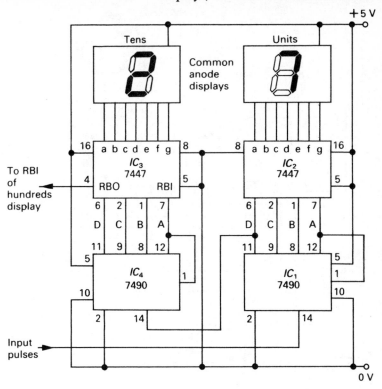

Fig. 11.9 Using TTL devices in a two-digit counter

they all appear to be on due to the eye's persistence of vision. A flashing light which occurs at a frequency above 20 to 30 Hz appears to the eye to 'run together' to give the effect of a continuous light. A multiplexed display illuminates all the appropriate segments of each digit for a fraction of a second, and immediately after does the same for next one. Fig. 11.10 shows how this is done for multiplexed 4-digit seven-segment display.

The scanning oscillator synchronises the two multiplexers, MUX1 and MUX2, which act like single-pole 4-way rotary switches. MUX1 takes the four outputs from the four BCD counters and supplies them as inputs to the BCD decoder/driver. At the instant that the BCD signal is routed to a particular digit, MUX2 energises that digit by connecting the cathodes of its seven LEDs to 0 V (assuming they are common-anode devices). The brightness of

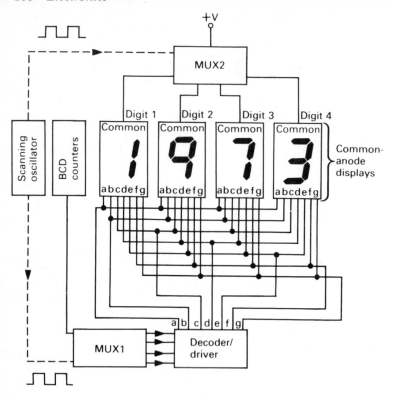

Fig. 11.10   Multiplexing a multidigit display

the display can be varied by varying the period during which each display is off. This is usually achieved by varying the mark-to-space ratio of the clock which operates the multiplexers. Since each segment LED is only illuminated for a short time, fairly high maximum currents can be passed through each segment, resulting in a bright display, yet one which requires, on average, less current than a non-multiplexed one. All the circuitry – counters, decoders, multiplexers, etc. – required to drive four or more digits in digital displays is readily available in integrated circuit packages.

12

Binary and Hexadecimal Arithmetic

12.1 Number systems

When we count things, we use the decimal number system. This number system uses the ten familiar digits 0 to 9 so that we can express the magnitude of any number we like. Thus the decimal number 273.16, two hundred and seventy three point one six, is made up from the ten digits by giving a decimal 'weighting' to each digit. The weighting depends on where the digit is in relation to the decimal point. Digits to the left of the decimal point have their weightings increased progressively by powers of ten. Digits to the right of the decimal point have their weightings reduced progressively by powers of ten.

The number 273.16 is made up of these decimal weightings as follows:

increasing powers of ten	digit	weight	digit × weighting	decimal number
	2	10^2	2×10^2	200
	7	10^1	7×10^1	70
	3	10^0	3×10^0	3
	1	10^{-1}	1×10^{-1}	.1
	6	10^{-2}	6×10^{-2}	.06
				273.16 (sum)

It is useful when looking at different number systems to distinguish this decimal number using the subscript '10', i.e. 273.16_{10}.

A binary digit, 0 or 1, is known as a **bit** which is probably a contraction of binary digit – or possibly it means just a small piece of information! In computing jargon, four bits, e.g. 1101, makes a

nibble, and eight bits, e.g. 11010010, makes a **byte**. The general name for a bit pattern is a **word** (though some people would have it called a gulp!). Bits are discussed in Chapter 9 where the digits 0 and 1 are the information of logic gates. And 4-bit words are handled by the digital counters described in Chapter 10.

A hexadecimal number is commonly used to simplify an 8-bit word (a byte used in many computer systems) into easily remembered two-character numbers. For example, the decimal number 119_{10} has a binary equivalent of 01110111_2. In hexadecimal this byte is simply 77_{16}. Note the use of the subscripts '2' and '16' to identify binary and hexadecimal numbers, respectively. The next section shows there are a few short cuts for converting numbers between decimal, binary and hexadecimal. The table below shows the binary and hexadecimal equivalents of the first twenty decimal numbers.

Decimal	Binary	Hexadecimal
0	0	0
1	1	1
2	10	2
3	11	3
4	100	4
5	101	5
6	110	6
7	111	7
8	1000	8
9	1001	9
10	1010	A
11	1011	B
12	1100	C
13	1101	D
14	1110	E
15	1111	F
16	10000	11
17	10001	12
18	10010	13
19	10011	14
20	10100	15

12.2 Converting between number systems

(a) Binary to decimal
We noted above that the decimal number 273.16_{10} is constructed as follows:

$$273.16_{10} = 2 \times 10^2 + 7 \times 10^1 + 3 \times 10^0 + 10^{-1} + 6 \times 10^{-2}$$

Now binary numbers have a base of 2 so the binary number 1101.01_2, for example, is constructed as follows:

$$
\begin{aligned}
1101.01_2 &= 1 \times 2^3 \quad + 1 \times 2^2 + 0 \times 2^1 + 1 \times 2^0 + 0 \times 2^{-1} + 1 \times 2^{-2}\\
&= 8 \qquad\quad + 4 \quad + 0 \quad + 1 \quad + 0 \qquad + 0.25\\
&= \qquad 13.25_{10}
\end{aligned}
$$

Thus to convert a binary number to a decimal number, it is only necessary to convert the powers of two to their decimal values and add the products. Alternatively, simply write the powers of two over each '1' bit and ignore those over the '0' bits.

(b) Decimal to binary
The short-cut way of doing this is to repeatedly divide the decimal number by two and record the remainder after each division. These remainders are either 1 or 0 and form the binary number. For example, suppose the number 202_{10} is to be converted to binary.

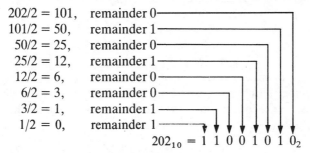

```
202/2 = 101,   remainder 0
101/2 = 50,    remainder 1
 50/2 = 25,    remainder 0
 25/2 = 12,    remainder 1
 12/2 = 6,     remainder 0
  6/2 = 3,     remainder 0
  3/2 = 1,     remainder 1
  1/2 = 0,     remainder 1
              202₁₀ = 1 1 0 0 1 0 1 0₂
```

If decimal fractions are to be converted, the rule is to multiply by two for digits on the right of the decimal point and to record the whole number integer. Thus the decimal number 0.3_{10} is converted as follows:

```
0.3 × 2 = 0.6,   integer = 0
0.6 × 2 = 1.2,   integer = 1
0.2 × 2 = 0.4,   integer = 0
0.4 × 2 = 0.8,   integer = 0
0.8 × 2 = 1.6,   integer = 1
              0.3₁₀ = .0 1 0 0 1₂
```

The multiplication is carried on until you have the required accuracy in the conversion.

(c) Hexadecimal to decimal

Since hexadecimal numbers have a base of 16, the decimal number is found by adding together the sum of the powers of 16. Thus, for the hexadecimal number $F7_{16}$:

$$F7_{16} = F \times 16^1 + 7 \times 16^0 = 15 \times 16^1 + 7 \times 16^0 = 247_{10}$$

(d) Decimal to hexadecimal

All that is required is to divide the decimal number by 16 repeatedly and record the remainder. For example:

$$59_{10}/16 = 3, \text{ remainder } 11 \ (B_{16})$$
$$3_{10}/16 = 0, \text{ remainder } \ \ 3$$
etc.
$$59_{10} = 3B_{16}$$

If the decimal number is fractional, e.g. 0.7_{10}, the digits to the right of the decimal point should be multiplied by sixteen and the result expressed as for the decimal-to-binary conversion. Thus

$$0.7_{10} \times 16 = 11.2, \quad \text{integer} = 11 \ (B_{16})$$
$$0.2_{10} \times 16 = 3.2, \quad \text{integer} = \ \ 3$$
etc.
$$0.7_{10} = 0.B3_{16}$$

(e) Binary to hexadecimal

If the binary number is a byte long, first divide it into two nibbles and then write down the hexadecimal equivalents of the two nibbles. Thus,

$$11100110_2 = 1110\,0110_2$$
$$= E6_{16}$$

If the binary number is not a byte long, the following examples will make the conversion clear:

$$0110_2 = 7_{16}$$
$$111011_2 = 0011\,1011 = 3B_{16}$$

(f) Hexadecimal to binary

The conversion is simply a matter of assigning the binary equivalent to the hexadecimal digit. Thus

$$F8_{16} = 1111\,1000_2$$
$$7D4_{16} = 0111\,1101\,0100_2$$

12.3 Binary arithmetic

First let's remind ourselves of how to add two decimal numbers, e.g. 239_{10} and 823_{10}:

Column	D	C	B	A
augend		2	3	9
addend		8	2	3
sum	1	0	6	2

Starting with the right hand column, the addition of 9 and 3 gives 12. This number is greater than base 10 of the decimal system so a '2' is recorded in the 'units' column A, and a '1' is recorded in the 'tens' column B. Next, the contents of column B are added, giving a total of 6 ('tens') which is less than the base 10 so there is no carry to column C. This procedure is repeated for the 'hundreds' column C.

Now the same rules can be applied to the addition of two binary numbers, e.g. 11010_2 and 10011_2. Thus:

Column	F	E	D	C	B	A
augend		1	1	0	1	0
addend		1	0	0	1	1
sum	1	0	1	1	0	1

Binary addition rules
$0 + 0 = 0$
$0 + 1 = 1$
$1 + 0 = 1$
$1 + 1 = 10$ (0 carry 1)
$1 + 1 + 1 = 10 + 1 = 11$

In column A, $0 + 1 = 1$ which is less than the base of 2, so there is no carry digit. Next for column B, $1 + 1 = 10_2$, since $1_2 + 1_2$ equals the base 2 and 0 is recorded in the 'twos' column and a carry digit goes to the '2' column C. Likewise the addition is carried to the remaining columns using the binary addition rules shown above.

When two decimal numbers are subtracted, the following procedure is used. Thus:

Column	D	C	B	A
minuend		3	5	9
subtrahend		1	8	7
difference		1	7	2

Starting with the right hand column A, the subtraction of 7 from 9 leaves 2. But in column B, 8 cannot be taken from 5 unless a carry digit is borrowed from the 'hundreds' column C, which reduces the digit 3 to 2. Subtracting 8 from 15 now gives 7 which is recorded in the difference column. Then for column C, 1 taken away from 2 (not 3) gives a difference of 1 in the 'hundreds' column C.

Similar rules are obeyed when subtracting two binary numbers. Thus the subtraction of 10010_2 from 11001_2:

Column	E	D	C	B	A
minuend	1	1	0	0	1
subtrahend	1	0	0	1	0
difference	0	1	0	1	1

Binary subtraction rules
$0 - 0 = 0$
$1 - 0 = 1$
$1 - 1 = 0$
$0 - 1 = 1$ and borrow 1
$10 - 1 = 1$

In column A, $1 - 0 = 1$. In column B, $0 - 1 = 1$ since a '2' (the base) is borrowed from column C. This borrowed '1' bit is 'put back' in the subtrahend of column C so that $1 - 0 = 1$, after borrowing a '2' from column D. When the borrowed '2' is put back in column D, we have $1 - 1 = 0$. Also the last column gives $1 - 1 = 0$.

12.4 Hexadecimal arithmetic

The addition of two hexadecimal numbers, e.g. $C6_{16}$ and $F8_{16}$ is as follows:

Column	C	B	A
augend		C	6
addend		F	8
sum	1	B	E

Decimal equivalents

$$C \times 16^1 + 6 \times 16^0 = 12 \times 16 + 6 = 198_{10}$$

$$F \times 16^1 + 8 \times 16^0 = 15 \times 16 + 8 = 248_{10}$$

$$\text{sum} = 446_{10}$$

$$1 \times 16^2 + B \times 16^1 + E \times 16^0 = 446_{10}$$

In column A, $6 + 8 = E_{16}$ which is less than the base sixteen so an E is recorded in the 'units' column A, and there is no carry digit to the 'sixteens' column B. Next for column B, $C + F = 1B$ since the addition is greater than the base sixteen and a carry digit is moved to the 'sixteen squared' column C. As shown above, the two hexadecimal numbers can be converted to decimal and the decimal numbers added to check the answer.

The subtraction of hexadecimal numbers is just as easy provided you remember that the base is sixteen. The following two examples show the technique.

Example 1

Column	C	B	A
minuend		E	6
subtrahend		8	2
difference		6	4

Example 2

Column	C	B	A
minuend		B	5
subtrahend		A	D
difference		0	8

In column A of example 1, $6 - 2 = 4_{16}$ as in decimal since we are subtracting equal digits in both systems. In column B, $E - 8 = 6_{16}$. In column A of example 2, $5 - D = 8_{16}$ after borrowing a '1' from column B. This is put back in column B to make the subtrahend $A + 1 = B_{16}$. Thus for column B, $B - B = 0_{16}$.

12.5 Binary adder circuits

Though Blaise Pascal's father was a mathematician, he forbade his son access to any books on mathematics. Instead he wanted him to study ancient languages. But Blaise so impressed his father with his understanding of geometry that his father gave in and let the boy study mathematics. In 1642 when he was nineteen, Blaise Pascal invented a calculating machine that, by means of cogged wheels, could add and subtract. This machine was the ancestor of the

modern mechanical cash register which has now given way to
electronic machines which perform arithmetic at lightning speed.

Calculators and microcomputers manipulate binary numbers
using circuits that add and subtract. In a computer, it is the
arithmetic and logic unit (ALU) inside its microprocessor that
performs the arithmetic. The most basic arithmetic unit is called the
half-adder. The half-adder does what we do mentally when we add
two binary digits; i.e. $0 + 0 = 0$, $1 + 0 = 1$, $1 + 1 = 10$. Fig. 12.1
shows one way of designing a circuit to add two single-bit binary
numbers. It is known as a half-adder and is made from an Exclusive-
OR gate and an AND gate. Since the circuit has two inputs, there
are 2^2, i.e. 4, combinations of inputs to consider and they are
summarised in the following table.

A B	sum $= A \oplus B$	carry $= A \cdot B$
0 0	0	0
0 1	1	0
1 0	1	0
1 1	0	1
bits added	XOR gate	AND gate

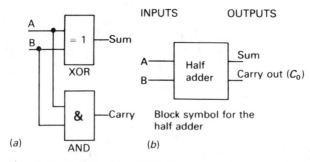

Fig. 12.1 A half-adder: (a) logic diagram; (b) circuit symbol

The 'sum' column shows why an Exclusive-OR gate (Section 9.3)
is necessary, while the 'carry' column justifies the use of an AND
gate. Thus the half-adder can only deal with the addition of the least
significant bit. Clearly we need a system that not only adds A and B

but also copes with the carry bit. Such a system is called a **full-adder**. Fig. 12.2 shows one way of constructing it from two half-adders and one OR gate. It accepts two binary digits, A and B, plus a carry-in bit, C_{in}. The 8 lines in the truth table below show all the additions the full-adder can do.

inputs	outputs			
carry-in, C_{in}	B	A	sum	carry-out, C_{out}
0	0	0	0	0
0	0	1	1	0
0	1	0	1	0
0	1	1	0	1
1	0	0	1	0
1	0	1	0	1
1	1	0	0	1
1	1	1	1	1
carry + A + B				

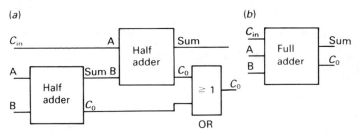

Fig. 12.2 A full-adder: (a) logic diagram; (b) circuit symbol

For instance, suppose A = 1, B = 1 and C = 0 as shown in Fig. 12.3a. The first half-adder has a sum of 0 and a carry of 1. The second half-adder has a sum of 0 and a carry of 0. Thus the final output is a sum of 0 and a carry of 1 as summarised in line 4 of the truth table. If A = 1, B = 1, and C = 1 as shown in Fig. 12.3b, line 8 of the truth table indicates that we get a sum of 1 and a carry of 1.

Two 4-bit binary numbers can be added together using three full-adders and one half-adder to produce a parallel binary adder. IC manufacturers produce several types of adders. The TTL 7483 and the CMOS 4008 integrated circuit devices are both 4-bit full-adders. These devices are shown in Fig. 12.4, and Fig. 12.5 shows a

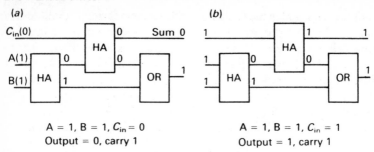

(a)

$C_{in}(0)$ — HA — Sum 0

A(1) — HA — OR — 1

B(1)

A = 1, B = 1, C_{in} = 0
Output = 0, carry 1

(b)

1 — HA — 1

1 — HA — OR — 1

1

A = 1, B = 1, C_{in} = 1
Output = 1, carry 1

Fig. 12.3 Using full-adders

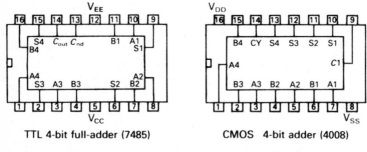

TTL 4-bit full-adder (7485)

CMOS 4-bit adder (4008)

Fig. 12.4 Two 4-bit integrated circuit full-adders

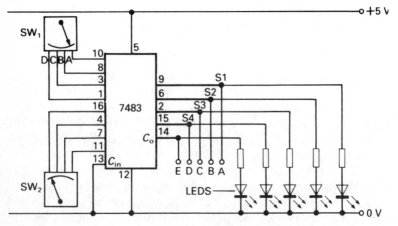

Fig. 12.5 A 4-bit full-adder using the 7483 device

practical 4-bit adder based on the 7483 device. Switches SW_1 and SW_2 are binary-coded rotary switches which select a number in the range 0 to 9 and provide the two 4-bit numbers to be added. The 7483 adds these two numbers to give a 5-bit result which can be observed on the LEDs, or on a seven-segment display (see Chapter 10). Thus when the two 4-bit numbers, 1111_2 and 1111_2 (decimal 15) are added, the result is 11110_2 (decimal 30). The most significant bit is available at the carry-out output (pin 14), and the first carry-in (pin 13) is connected to 0 V. Several 7483s can be used to add longer binary words.

13

Memories and Silicon Chips

13.1 Introduction

Just like human memory, electronic memory is simply any device that stores information for future use. This information is stored in an electronic memory as a collection of binary digits (or bits), i.e. 1s and 0s. Memory is therefore a feature of digital systems and not of analogue systems. The information, or data, is usually stored in digital memory as groups of 8-bit words (known as bytes), or 16-bit words. Microcomputers, calculators, electronic games and an increasing number of other digital systems make extensive use of memory.

The way memory is used in microcomputers is shown in the systems diagram of Fig. 13.1. Here data flows between the memory

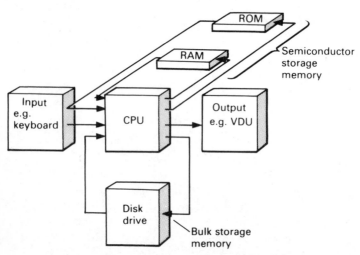

Fig. 13.1 The basic microcomputer system

devices and a central processor unit or CPU (also called a micro-processor), and between input and output devices. In a computer system such as this, memory is categorised either as random-access memory (RAM), or as read-only memory (ROM). Both types of memory may be made from thousands of transistors formed as an integrated circuit on a small chip of silicon as explained in Section 13.12. In addition to RAM and ROM semiconductor-storage memories, data may be stored outside the computer system as bulk-storage memory which includes magnetic tape and disk.

13.2 Random-access memories (RAMs)

The RAM-type of memory can be made to 'learn', a process called 'writing information into it'. And the data which it remembers can be recalled at any time, a process called 'reading data from it'. Some people prefer to call the RAM a 'readily-alterable memory' since it is a device whose contents can be altered quickly and easily. With very few exceptions, RAMs lose their contents when the power is removed and are thus known as 'volatile' memory devices. All microcomputers use RAM to store data and programs written (or loaded) into it from a keyboard, or from an external store of data such as magnetic tape or disk.

RAMs are often described in terms of the number of bits, i.e. 1s and 0s, of data that they can hold, or in terms of the number of data words, i.e. groups of bits, they can hold. Thus a 16 384 bit RAM can hold 16 384 1s and 0s. This data could be arranged as 16 384 1-bit words, 4096 4-bit words, or 2048 8-bit words. Semiconductor memories vary in size, e.g. 4K, 64K, 128K, etc. Here we are using K defined as

$$K = 2^{10} = 1024.$$

Thus a 16K memory has a storage capacity of $16 \times 1024 = 16\,384$ words, a 128K memory of 131 072 words, and so on.

There are two main members of the RAM family: static RAM and dynamic RAM. The essential difference between them is the way in which bits are stored in the RAM chips. In a static RAM, the bits of data are written in the RAM just once and then left until the data is read or changed. In a dynamic RAM, the bits of data are repeatedly rewritten in the RAM to ensure that the data is not forgotten.

13.3 Static RAMs

Flip-flops (Chapter 10) are the basic memory cells in static RAMs. Each flip-flop is based on either two bipolar transistors, as shown in Fig. 13.2, or on two metal-oxide semiconductor field-effect transistors (MOSFETs). As many of these memory cells are needed as there are bits to be remembered. Thus in a 16K-bit static memory, there are 16 384 flip-flops, i.e. 32 768 transistors. All these transistors would be accommodated on a single silicon chip about 4 mm square.

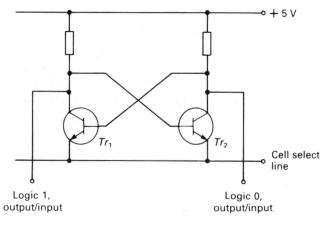

Fig. 13.2 The basic memory cell in a static RAM

The 7489 TTL device shown in Fig. 13.3 is an example of a simple static RAM. It has 64 memory cells, each cell capable of holding a single bit of data. The cells are organised into locations, and each location is capable of holding a 4-bit word. Thus the 7489 is capable of storing sixteen 4-bit words, i.e. four memory cells are used at each location. Each location is identified by a unique 4-bit address so that data can be written into or read from these locations. Note that the number of words stored in memory determines the size of the address word. Thus the number of address SELECT lines (pins 1, 15, 14, and 13) is four since $2^4 = 16$. There are four data input lines (pins 4, 6, 10, 12) on which data is placed to be stored in memory. And there are four data output lines on which data is read out from the RAM.

Fig. 13.4 shows a model computer memory based on the 7489.

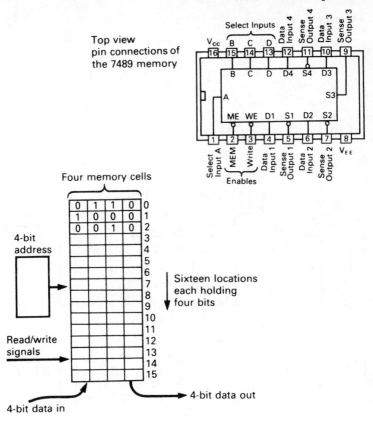

Fig. 13.3 Memory organisation of the 7489 static RAM

The 4-bit data latches, IC_1, IC_2 and IC_3, enable the memory to be addressed, and data to be transferred to and from the memory. Each of these three ICs contains four flip-flops which are 'loaded' with data when its pins 4 and 13 go from LOW to HIGH. Data is loaded into each of the 16 locations and read back from them as follows.

(i) To write data into the memory. First set SW_{12} to the 'read memory' position. (In a practical circuit, SW_{12} should be a momentary action push-button switch to avoid writing data into memory inadvertently.) Next select the 4-bit address (DCBA) of the chosen location by setting switches SW_5 to SW_8: if one of these switches is left open, the bit is set to 0 by the resistors; if closed the bit is set to 1.

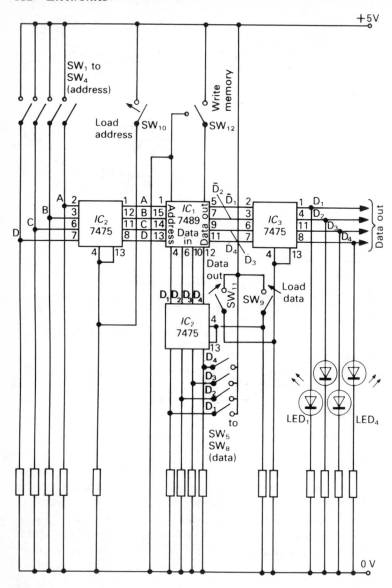

Fig. 13.4 A model computer memory using the 7489 static RAM

Power supply connections:
+ 5 V to pin 16 of IC_1 and pin 5 of IC_2, IC_3 and IC_4
0 V to pin 8 of IC_1 and pin 12 of IC_2, IC_3 and IC_4 All resistor values = 1K

Now 'strobe' this address into IC_2 by SW_{10}. Next use switches SW_5 to SW_8 to select the 4-bit data ($D_4D_3D_2D_1$) to be loaded into the address location. 'Strobe' this data into IC_1 by means of SW_9. Now operate SW_{12} so that this data is written into memory. Repeat the above sequence to write data into the other locations.

(ii) To read data from memory. Ensure that SW_{12} is in the 'read memory' state. Select the required address location from which data is to be read by setting switches SW_1 to SW_4, and load it into memory as before. This data is now available on pins 5, 7, 9 and 11 of IC_1. Operate switch SW_{11} so that data from memory is transferred to the output of IC_3, and can be read on the LEDs.

13.4 Dynamic RAMs

Unlike a static RAM which holds its data until 'told' to change it, a dynamic RAM continually needs to have its data refreshed. Dynamic RAMs (also called DRAMs) are based on metal-oxide semiconductor field-effect transistors (MOSFETs – see Chapter 8). Bits of data are stored in dynamic RAMs as small packets of charge, rather than as voltage levels as in the static RAM. This has the advantage that the power consumption of MOSFET memory circuits is very low.

As shown in Fig. 13.5, each memory cell in a DRAM is a very simple circuit and comprises a small capacitor and a single n-channel MOSFET which is switched on to read this charge. Each cell holds a 1 bit as a tiny electrical charge of about 10^{-15} coulombs. Though

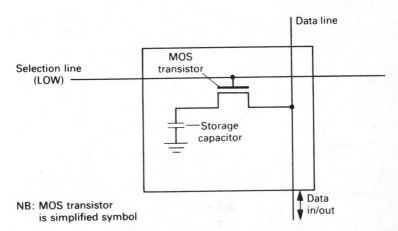

Fig. 13.5 The basic memory cell in a dynamic RAM

tiny, this charge still amounts to about 5000 electrons! However, the charge on the capacitor tends to leak away and extensive 'refresh' circuitry is needed to keep the charge 'topped up'. The additional electronics required to ensure that a dynamic RAM's memory doesn't forget what's in it adds to the cost and complexity of dynamic RAMs. However, the newer dynamic RAMs have refresh circuitry on the chip with the memory cells.

Dynamic RAMs tend to be cheaper than static RAMs for large-capacity memory devices. This is because of the smaller size of the dynamic RAM cell as opposed to the static RAM cell, as it is based on one transistor instead of two. The first 1K-bit DRAM was introduced in the early 1970s, and since then the number of cells on a memory chip has doubled every year, culminating in the latest 1M-bit devices, and the proposed 4M-bit memories by 1989. Today the market for memories accounts for over half the total market of integrated circuits.

13.5 Read-only memories (ROMs)

The problem with a random-access memory is that its memory is volatile, i.e. it loses all its data when the power supply is switched off. A non-volatile memory is a permanent memory that never forgets its data. One type of non-volatile memory is the read-only memory (ROM). A ROM has a pattern of 0s and 1s imprinted in its memory by the manufacturer. It is not possible to write new information into a ROM, which is why it is called a 'read-only' memory.

The organisation of data in a ROM is similar to that of a RAM. Thus a 256-bit ROM might be organised as 32 words each 8-bits long; a 1024-bit (1K-bit) ROM might be organised as a 256 × 4-bit memory, and so on. Fig. 13.6 shows the 7488 256-bit ROM organised as 32 × 8-bit words. Any one of these 8-bit words may be addressed through the five SELECT lines which identify the location of each 8-bit word. When pin 15, the MEMORY ENABLE pin, is taken LOW, the word stored in the location appears on the eight output lines.

A ROM may be regarded as the 'reference library' of the computer world. For example, microcomputers have an integrated circuit ROM which stores instructions such as the language and graphics symbols the computer uses. A typical ROM for a micro-computer has 8 kilobytes of memory (1 byte is an 8-bit word). Thus it stores 65 536 bits of data on a single chip which is generally

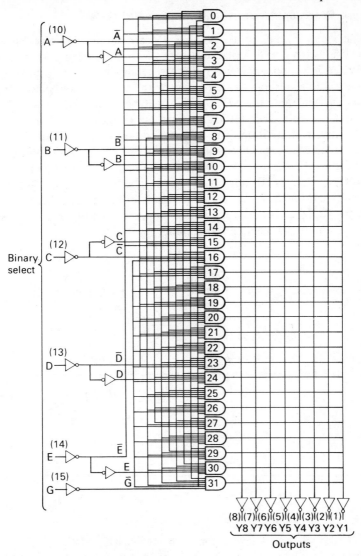

Fig. 13.6 An extreme simplification of how memory is organised in the TTL 7488 256-bit ROM (pin numbers are shown in brackets)

contained in a 28-pin d.i.l. package. Through these 28 pins, the microcomputer is able to select and read any one of the 8192 locations in which the bytes of data are stored.

13.6 Programmable read-only memories (PROMs)

The memory cells in a bipolar ROM are simply npn transistors which have a fuse link placed in series with the emitter as shown in Fig. 13.7. Suppose a decoder addresses transistor Tr_1, which is one of eight in a stored 8-bit word, say. With the fuse link in place, Tr_1 will be 'on' and the SENSE line will be pulled HIGH by the addressed cell. This HIGH forward-biases the buffer transistor, Tr_2, thus turning it on. The collector voltage of Tr_2 is then LOW, i.e. at logic 0. However, if the fuse link is open, the transistor is 'off' and the SENSE line will be pulled LOW. Hence the collector voltage of Tr_2 is HIGH, i.e. at logic 1. The fuse links are made of some material such as titanium–tungsten or nichrome. These transistor ROM cells are programmed by one of two processes, mask programming and field programming.

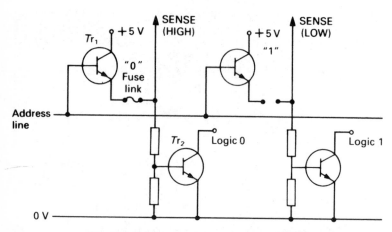

Fig. 13.7   The use of a fuse link in a PROM

Mask programming is accomplished when a ROM is made. If a '1' is to be stored at a particular memory location, the fuse link of the transistor at that location is simply not connected. The second method of programming a ROM is left to the user. In a **field-programmable** ROM, i.e. a PROM, the user stores a '1' in a cell by addressing the cell and applying a high current pulse (approximately 30 mA) to it which 'blows the fuse'. This once-only programming is done with a device called a PROM burner. To read the contents of a particular cell, the cell is addressed by an address decoder that is usually an integral part of the PROM chip. ROMs are used for dedicated applications such as controlling the actions of an industrial robot, or providing a microcomputer with a word-processing capability.

Both mask and programmable ROMs are rendered useless if there is just one incorrect bit of data stored in the chip. Fortunately another type of ROM, the EPROM (or erasable–programmable ROM) has a memory which can be erased and reprogrammed many times. A typical EPROM is shown in Fig. 13.8 and is based on MOSFET transistors. The quartz window over the top of the chip allows ultraviolet light to erase any stored data in the chip. This usually takes about 10 minutes. A fresh set of instructions is written in the EPROM using an EPROM programmer which re-establishes a 'packet' of charge in selected MOSFET transistors. The charge remains there indefinitely unless erased by UV light.

Fig. 13.8   A typical EPROM showing the quartz window that allows UV light to reach the chip

13.7 Magnetic bubble memories (MBMs)

Bubble memories store bits of data on small magnetised domains called 'bubbles'. The domains move in a thin film of magnetic garnet which is deposited on a non-magnetic garnet substrate. It is possible to annihilate bubbles to clear the store, generate bubbles in order to write new data, and replicate bubbles for reading out the data. Stored data can be accessed by moving the bubbles along a closed track called a 'major loop' which links write, read and erase stations as shown in Fig. 13.9.

The track along which the bubbles are moved is defined by patterns of permalloy deposited on a thin magnetic film grown on to a substrate of gadolinium-gallium-garnet. The bubbles are created by electrical signals that circulate in very small conducting loops just above the film. They are made to move under the control of a rotating magnetic field. Bubbles are detected when they pass under the permalloy strips deposited on the bubble-bearing film. The magnetisation of the strips, and hence their resistance, changes due to the presence of a bubble (which represents a 1). The absence of a bubble represents a 0.

Data is transferred from the major loop to minor loops for storage; it is accessed by transferring data from the minor loops after these have been rotated so that the required data is next to the major loop. By using a number of minor loops, the time to access

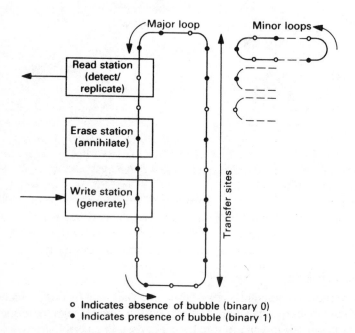

○ Indicates absence of bubble (binary 0)
● Indicates presence of bubble (binary 1)

Fig. 13.9 A simplification of the operation of a bubble memory

data can be reduced considerably since one bit from each of the minor loops can be transferred in parallel to the major loop.

The main advantage of an MBM is that it is non-volatile (like a ROM) since the bubbles do not disappear when the power supply is switched off. However, unlike a ROM an MBM can have data written into it at any time. MBMs are quiet and very reliable since they have no moving parts. And since an MBM can store over a million bits in a very small volume, it is suitable for a whole range of microprocessor-based products such as word-processors, data loggers and microcomputers.

13.8 Floppy disks

Another type of bulk storage device on which large amounts of data can be stored is the floppy disk (also called diskette, or mini-disk). The data is written to and read from floppy disks using a disk drive – see Fig. 13.10. There are several sizes of floppy disks but the one commonly used with microcomputers is 5.25 inches across. It is made of flexible Mylar film which is coated with a thin magnetic film on which binary data is stored in the form of tiny magnetised regions. Recently, a 3.5 inch disk has been introduced.

As shown in Fig. 13.10a, a floppy disk is enclosed within a plastic jacket. There is a round hole in the middle of the jacket which enables the hub of the disk drive to clamp on the disk and rotate it at 300 rev/min. Data is stored on and retrieved from the disk by the read/write head of the disk drive which touches the spinning disk through the oblong hole at the edge of the disk. Note that since the read/write head can move across the surface of the disk, the floppy disk (in common with the compact disk used for storing audio information) is a **random-access** device. This means that the read/write head of the disk drive (and the laser head on a compact disk) can jump to any location on the disk. Thus the data on a floppy disk can be accessed much faster than data on a magnetic tape. The tape player has to sequence through a lot of data on the tape before locating a certain section, i.e. it is a **sequential-access** device.

The data on a floppy disk is organised into concentric circles called **tracks** (Fig. 13.10b). There may be forty tracks on a disk and each track is divided into ten sectors. Each sector is further divided into 256 bytes of data where one byte (an 8-bit word) represents one stored character. A single-sided, single-density disk has 40 tracks on one side only; a double-sided, double-density disk has 80 tracks on

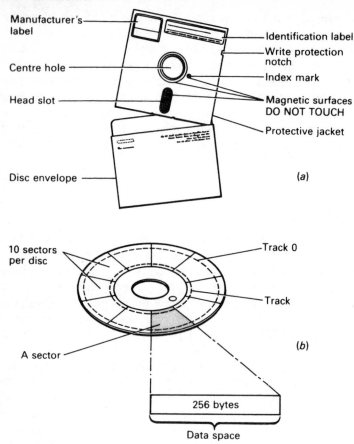

Fig. 13.10 A floppy disk: (a) its appearance; (b) organisation of its memory

each side. Thus a single-sided single density 5.25 inch floppy disk can store

$$40 \text{ tracks} \times 10 \text{ sectors} \times 256 \text{ bytes} = 102\,400 \text{ bytes}$$

The disk filing system of a computer automatically records the location of data, e.g. programs, on a disk. The first two sectors on track zero of a disk are reserved for this purpose – they hold the 'catalogue' of the disk. Whenever a piece of data is required from the disk, the disk filing system first reads the catalogue to find out which tracks and sectors are used to store this data. Clearly, before a floppy disk can be used, it has to have the tracks and sectors created on the disk so that the disk filing system can find the data required. This process is called 'formatting' a disk and includes setting up the tracks and sectors and creating the catalogue. A formatting routine is usually provided with a microcomputer to enable floppy disks to be formatted, but it should be remembered that there is no standard way of formatting a floppy disk.

13.9 Access time of a memory

The access time for a particular memory is the time it takes to locate and deliver a piece of data in the memory. RAMs and ROMs (semiconductor storage memories) have faster access times than disks, tapes and bubble memories (bulk-storage memories). Thus RAMs have an access time of less than 1 μs; e.g. the 6116 16K-bit MOSFET static RAM has an access time between 100 ns and 250 ns. Bipolar RAMs have faster access times than this. ROMs have slower access times than RAMs. Of the bulk storage devices, the floppy disk (a parallel-access memory) has an access time of about 30 ms. Slower still is magnetic tape (a serial-access device) with an access time of several seconds or even minutes. MBMs have access times of between 10 and 100 μs. But a short access time is not all-important and has to be traded against the cost per bit stored in a memory.

13.10 Gallium arsenide

Gallium arsenide (GaAs) is a crystalline substance which, like silicon, is used to make diodes, transistors and integrated circuits. Gallium arsenide's main claim to fame is that semiconductor devices made from it conduct electricity five times faster than those made from silicon. And this property makes gallium arsenide an interesting material to weapons manufacturers since data in computer memories made from GaAs can be accessed more quickly than memories made from silicon. This means that missiles and other weapons can respond rapidly to sensing and control circuits.

However, there are a few drawbacks to the use of GaAs, one of which is that the two elements gallium and arsenic from which it is made are in short supply, mainly being found as impurities in aluminium and copper ores. On the other hand, silicon is plentiful, being found in silicates such as sand. The cost of GaAs is about thirty times that of silicon, and this is aggravated by the fact that about 90% of GaAs chips are rejected after production. Furthermore, it is not as easy to make integrated circuits from GaAs as it is from silicon since it does not form a protective layer of oxide to resist the diffusion of dopants during the process of photolithography – see Section 13.12.

It is therefore unlikely that there will be a rapid rise in the use of GaAs-based semiconductors in the near future except for specialist applications where cost is not a major consideration, e.g. ballistic missile development and the USA's 'Star Wars' programme, the Strategic Defence Initiative (SDI). But manufacturers of GaAs have turned to space for help in overcoming the problems and costs of producing pure GaAs. A number of countries are planning to produce pure crystalline GaAs in the extremely low vacuum and zero gravity in laboratories aboard orbiting space stations.

13.11 Silicon chips and Moore's law

An integrated circuit comprises microscopically small transistors, diodes, resistors and other components connected together on a 2 to 5 mm square chip of silicon. The circuit produced generally has all the components necessary for the chip to function as an amplifier, or an analogue-to-digital converter, or a memory, and so on. The components making up integrated circuits are the smallest man-made objects ever created; and the trend is towards producing even smaller components on a silicon chip.

In the 1960s, George Moore, founder of the Intel Corporation in the USA, said that the number of components that would be integrated on a single silicon chip would double every year (Fig. 13.11). To the present day, Moore's law (as it came to be known) has been found to hold good. In Moore's day, ICs were made from a few tens of components. By the late 1960s, the annual doubling effect had led to several hundred components being integrated on a silicon chip (known as medium scale integration – MSI). During the 1970s, component counts per chip had reached several hundred thousand (known as large scale integration – LSI). In the 1980s, microprocessor and memory chips now have close to a million components on a single silicon chip, and this is called very large scale integration – VLSI. The small size of components of a silicon chip is illustrated by a simple analogy: if a present-day 5 mm square memory chip were the size of the UK, the smallest component would be the size of a tennis court.

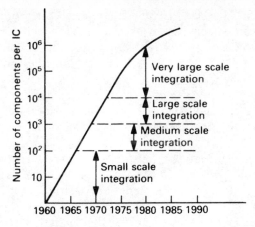

Fig. 13.11 Graph showing progressive increase in chip complexity

13.12 Making a silicon chip

The process of putting several hundred thousand transistors on a silicon chip a few millimetres square is not easy. A technique called **photolithography** is at the heart of the process. Photolithography uses photographic techniques and chemicals to etch a minutely detailed pattern on the surface of a silicon chip. Each stage in the process involves the use of photographically-prepared plates called photomasks. Each photomask holds a particular pattern identifying individual transistors, conducting pathways, etc. Photomasks are produced by the photographic reduction of a much larger pattern. A photomask is placed over a thin layer of photoresist covering the surface of the silicon. Ultraviolet light shone on the photomask passes through the clear areas but is stopped by the opaque areas. According to the type of photoresist used, either the exposed or the unexposed photoresist can be dissolved away using chemicals to leave a pattern of lines and holes. This pattern enables transistors to be formed in the silicon, and aluminium interconnections to be made between them.

The process of making a silicon chip begins with a 50 to 150 mm cylinder-shaped single crystal of pure silicon (or gallium arsenide) known as a boule or ingot (Fig. 13.12). The ingot is obtained by slowly pulling the growing crystal from a bath of pure molten silicon. It is then cut up into thin slices known as wafers, about the size of a beer mat and half as thick.

The wafers are then passed through an oven containing gases

Fig. 13.12   A single crystal of pure gallium arsenide and a
polished slice of it
Courtesy: Cambridge Instruments Ltd

heated to about 1200 °C. The gases diffuse into each wafer to give it
the properties of a p-type or an n-type semiconductor (Chapter 2)
depending on the gas used – a process known as epitaxial growth.
The wafers are then ready to have integrated circuits formed on
them by a complex process which involves masking, etching and
diffusion.

Fig. 13.13 shows the several stages required to produce openings
in the surface of the silicon through which gases are diffused to
create transistors. The first stage involves heating the wafer to about
1000 °C in a stream of oxygen so that a thin layer of silicon dioxide is
formed over the whole surface of the wafer.

In the next stage, a thin layer of a light-sensitive emulsion
(photoresist) is spread over the layer of silicon dioxide. A photographic plate (the photomask) is placed over the top of the emulsion. The photomask contains a pattern of dots in microscopic detail
which are to become holes in the silicon dioxide layer. A single mask

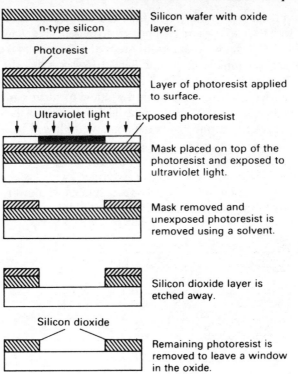

Silicon wafer with oxide layer.

Layer of photoresist applied to surface.

Mask placed on top of the photoresist and exposed to ultraviolet light.

Mask removed and unexposed photoresist is removed using a solvent.

Silicon dioxide layer is etched away.

Remaining photoresist is removed to leave a window in the oxide.

Fig. 13.13 Steps in the formation of a window in the silicon dioxide surface of a silicon chip

holds the pattern for several hundred integrated circuits for each wafer.

In the third stage, the mask is exposed to ultraviolet light. Where the mask is transparent, the light passes through and chemically changes the photoresist underneath so that it hardens. The unexposed photoresist can easily be removed with a suitable solvent (fourth stage).

In the fifth stage, the silicon wafer is immersed in another solvent which removes the silicon dioxide from the unexposed areas. The wafer now has a thin surface layer of silicon dioxide in which there are a large number of minute 'windows'. It is through these windows that gases are allowed to pass into the epitaxial silicon layer underneath to form transistors. In the production of complete silicon chips on a wafer, the formation of a silicon dioxide layer, followed by masking and etching, has to be repeated many times.

Fig. 13.14 shows the various steps required to make an npn transistor on a silicon chip. And Fig. 13.15 shows that similar steps are required to create a single n-channel MOSFET. First a gas is selected which diffuses through a window to form a p-type base region in the n-type silicon epitaxial layer. Next a fresh silicon dioxide layer is formed over the window, followed by a stage of masking and etching to create a second smaller window. Through this window a gas diffuses to form the n-type emitter region. Another layer of silicon dioxide is formed over this window followed by masking and etching to create smaller windows for making contacts to the base and emitter regions. These contacts are made

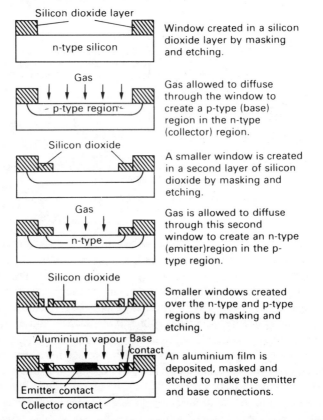

Fig. 13.14 Steps in the formation of a single npn transistor on a silicon chip

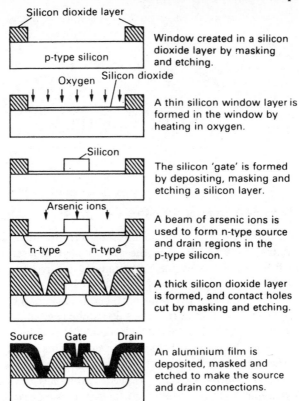

Window created in a silicon dioxide layer by masking and etching.

A thin silicon window layer is formed in the window by heating in oxygen.

The silicon 'gate' is formed by depositing, masking and etching a silicon layer.

A beam of arsenic ions is used to form n-type source and drain regions in the p-type silicon.

A thick silicon dioxide layer is formed, and contact holes cut by masking and etching.

An aluminium film is deposited, masked and etched to make the source and drain connections.

Fig. 13.15   Steps in the formation of a single n-channel MOSFET transistor on a silicon chip

by depositing aluminium in vapour form. In the final stages of making an integrated circuit, vapourised aluminium is allowed to form a thin layer of aluminium over the entire surface of the silicon chip. This thin layer is cut into a pattern of conducting paths using the techniques of masking and etching. When all the integrated circuits have been formed in this way, the wafer is cut up into individual chips, checked and packaged in a form which can be used by the circuit designer. The IC package is usually the familiar dual-in-line version.

14

Control Systems

14.1 The nature of control systems

Control engineering is a vast field and ranges from simple thermostatic control systems for, say, the control of temperature in a tropical fish tank, to advanced position-control systems aboard spacecraft that explore the solar system. But whatever their level of sophistication, all control systems have certain common basic features. The simplest form of control is **open-loop control**. Fig. 14.1a shows its three basic elements. The building block called 'desired output' is what the user of the control system wants as the 'actual output' shown by building block 3. Building block 2, the 'controller', makes the output possible after the input has been set. A typical example of an open-loop control system is a domestic light dimmer switch. The light level is selected by the amount a control knob is turned. But if the light dims because of a partial power failure, there is no feedback between the output (the amount of light produced) and the input (the control setting) to maintain the light level. An open-loop control system has no 'feedback' to enable it to make changes to the actual output once the input has been set.

But, having set the dimmer switch you could intervene to bring the light level back to what you wanted. Now you have taken action and provided feedback, the dimmer switch has become a **closed-loop control** system. The system is a closed-loop control system as long as you, the 'human operator', remain on duty. In automatic closed-loop control systems, the human operator is replaced by a monitoring device which improves the action of the human operator, e.g. the automatic device can't go to sleep! Fig. 14.1b shows the basic elements of a closed-loop control system. The system includes a transducer for monitoring the prevailing state of the actual output and converting it into a form similar to the signal

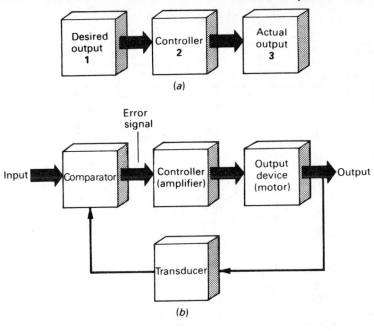

Fig. 14.1 Two basic types of control system: (a) open-loop control; (b) closed-loop control

representing the desired output: these two signals are compared to produce a difference, or error, signal which is then used to control the system. Thus closed-loop control systems are 'error-actuated'. Modern control systems use a variety of electrical transducers for producing the feedback signal. For example, temperature can be monitored with a thermistor (Section 5.6) or with a thermocouple (Section 15.2). Position may be monitored with a variable resistor, e.g. a potentiometer (Section 5.3), and load with a strain gauge (Section 5.8).

14.2 The design of a thermostat

The simple thermostat circuit shown in Fig. 14.2 is an example of a closed-loop control system. It comprises three parts: a thermistor temperature sensor, Th_1, which is part of a voltage divider; a comparator based on an integrated circuit, IC_1, called an operational amplifier; a single transistor amplifier, Tr_1, which opens and closes the relay contacts to control the power to the lamp. The

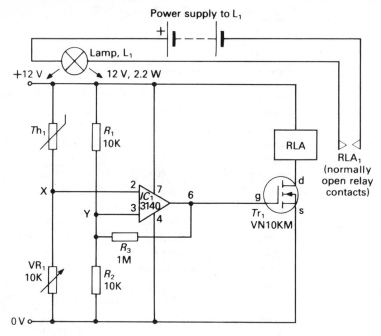

Fig. 14.2 A simple thermostat

detailed operation of an operational amplifier (op. amp.) is described in Section 14.5. In this circuit, the op. amp. compares the voltage set on pin 3 with that on pin 2. It produces a positive voltage to switch the transistor on when the voltage on pin 2 is less than that on pin 3.

In this simple circuit, the heat is generated by a 12 V filament lamp. In use, the lamp could heat the air in a small box which houses tropical insects, for example. The thermistor senses the temperature in the box. Suppose VR_1 is set so that the lamp just switches on, i.e. the output voltage from the comparator is positive. In this case, the voltage on pin 2 of the comparator is slightly lower than the voltage on pin 3. As the heat from the lamp warms the air inside the box, the resistance of the thermistor falls. When this fall makes the voltage at pin 2 rise above that on pin 3, the output voltage from the comparator falls to 0 V and the heater is switched off. Thus the temperature of the box rises or falls to a value determined by the reference voltage set on pin 3 by VR_1. A scale could be fitted to the spindle of VR_1 so that the thermostat could be used to stabilise selected temperatures. The purpose of resistor R_3 is explained in Section 14.6.

14.3 The design of a motor speed controller

Fig. 14.3 shows a simple open-loop control system for controlling the rotational speed of a common permanent magnet d.c. motor. The speed of the motor is set by the potentiometer VR_1 which controls the base–emitter current of the npn transistor Tr_1. The amplified current produced by Tr_1 flows through the motor's armature coil which is connected between the emitter of Tr_1 and 0 V. As VR_1 is adjusted to raise the base–emitter voltage, V_{be}, above 0.6 V, the motor begins to rotate as Tr_1 amplifies the base–emitter current. But when the motor rotates, a back e.m.f. is generated by the armature coil. This raises the voltage at the emitter of Tr_1 and tends to turn the transistor off. But this back e.m.f. cannot turn the transistor fully off since this would reduce the back e.m.f. and increase the current flowing through the transistor. Thus the motor settles down to a speed determined by the setting of VR_1 which dictates the base–emitter current of Tr_1. Any loading of the motor reduces its speed, which decreases the back e.m.f. and lowers the emitter voltage so allowing more current to flow through the motor. Thus the speed of the motor tends to stabilise for a given setting of VR_1. But this is still an open-loop control system since no information about the output, i.e. the motor's speed, is available to control the input. The way the back e.m.f. stabilises the action of this circuit

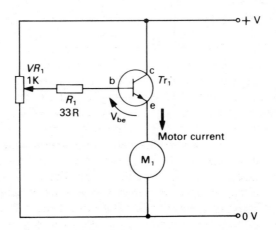

Fig. 14.3   A simple motor speed controller

by reducing the effect of the input is an example of negative feedback – see Section 15.5 for a fuller explanation of negative feedback.

14.4 The design of a servosystem

A servosystem is an electromechanical device for the precise positioning of something, e.g. the positioning of a laser beam over the tracks in a compact disk player. Thus a servosystem uses transducers to continuously monitor the position of the output, and provides corrective action to ensure that the output behaves correctly. A servosystem is an example of a closed-loop control system. Fig. 14.4 shows a simple but useful example of a servosystem based on an integrated circuit operational amplifier (or op. amp.), IC_1. The circuit is designed to position an output shaft (attached to the d.c. motor M_1) to a desired angle by the setting of the potentiometer VR_1. This automatic setting of the output according to a preset input is provided by the mechanical link of a second potentiometer,

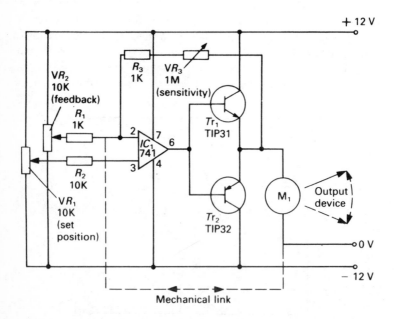

Fig. 14.4 A simple servosystem

VR$_2$, attached to the motor shaft. Thus this servosystem could be used for the remote rotation of a roof-top aerial to obtain optimum reception of radio signals.

VR$_3$ is used to set the sensitivity of the servosystem by altering the voltage gain of the op. amp. – see Section 15.5. If the gain is too high, the servosystem responds too sharply to any change in VR$_1$, the 'set position' potentiometer. This usually means that the output shaft will overshoot the required setting and oscillate about a mean position. These oscillations may be 'damped', i.e. die down after a few oscillations, or continue – a condition known as 'hunting'. If the voltage gain is set too low, the output will follow the setting of the input potentiometer very sluggishly. For optimum response, the setting of VR$_3$ must be such that the output reaches the desired position quickly and without overshoot. The transistors shown are suitable for controlling currents up to about 3 A using heat sinks bolted to them – see Chapter 8. Since the op. amp. is operated from a dual power supply, its output voltage can be positive or negative with respect to 0 V. When it is positive, the npn transistor Tr$_1$ switches on and current flows through the motor one way; when it is negative, the pnp transistor Tr$_2$ switches on and current flows through the motor the opposite way.

Whether the output voltage is positive or negative depends on the difference of voltage between pins 2 and 3 of the op. amp. This difference is the error signal, which is reduced to zero by the feedback provided by the mechanical link between the output and the input. This action is called negative feedback and is explained more fully in Section 15.5. When the error signal equals 0 V, both transistors are switched off and the motor is stationary.

14.5 The basic properties of op. amps.

The thermostat and servosystem described in the preceding sections were based on an integrated circuit op. amp. (or operational amplifier). In this design, the op. amp. was used to compare a reference voltage set on one of its two inputs with a varying voltage on its other input. When the varying voltage was slightly higher or lower than the reference voltage, the output of the op. amp. changed abruptly. This abrupt change in output voltage is used to switch a relay on or off to form the basis of a simple thermal control system. In Chapter 15, you will see how op. amps. are used in instrumentation systems to amplify small differences of voltage so that temperature measurements can be made. The op. amp. is such an important integrated circuit in control systems and instrumentation systems that a few words are needed to explain what it does, and what its specially useful characteristics are.

The appearance of one type of op. amp. is shown in Fig. 14.5a. This is an 8-pin dual-in-line (d.i.l.) package. Fig. 14.5b shows how the pins are numbered looking at the package from the top. The general connections to the op. amp. are shown in Fig. 14.5c. Note its circuit symbol is a triangle, an arrow head that shows the direction in which signal processing takes place. The op. amp. has two inputs (pins 2 and 3), one output (pin 6) and two power supply connections (pins 4 and 7). An op. amp. can be operated with one or two power supplies as shown in Fig. 14.6. If it uses a single power supply, pin 4 is connected to 0 V and pin 7 to $+V$. Input and output signals are measured with respect to 0 V. With two power supplies, pin 4 is connected to the $-V$ terminal of one supply, pin 7 is connected to the $+V$ of the other supply, and the common connection of the two supplies provides the 0 V supply line. Now input and output voltages are measured with respect to this common connection.

But why does the op. amp. have two inputs? Fig. 14.5c shows that one input (pin 2) has an input voltage of V_2 on it, and the second input (pin 3) has an input voltage of V_1 on it. What does the op.

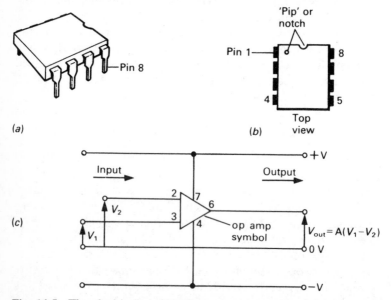

Fig. 14.5 The dual-in-line (d.i.l.) op. amp. package: (a) its general appearance; (b) its pin identification; (c) its circuit symbol

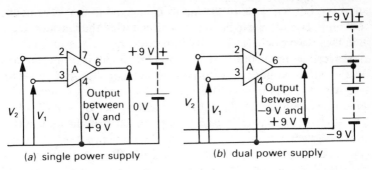

Fig. 14.6 Power supply connections to the 8-pin d.i.l. version of an op. amp.

amp. do with these two input voltages? It amplifies the difference between them so that the output signal V_{out} is given by the following 'op. amp. equation':

$$V_{out} = A(V_1 - V_2) = A \times V_{in}$$

where $V_1 - V_2$ is the input voltage, V_{in}, and A is known as the voltage gain of the op. amp. A is the number of times the output voltage is greater than the input voltage, and it is very high. Most modern integrated circuit op. amps. have voltage gains in excess of 100 000. Note that an op. amp. is oblivious to the actual values of the voltages on its two inputs – it only 'sees' the difference between them. The high voltage gain has to be tamed before it is of any use in the instrumentation systems described in Chapter 15. But control systems are designed to make use of this high gain.

One of the op. amp's two inputs (pin 2) is called the inverting input and the other (pin 3) the noninverting input. What do these names mean? Fig. 14.7 shows an op. amp. wired up with two power

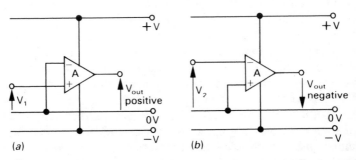

Fig. 14.7 The action of (a) the noninverting input and (b) the inverting input of an op. amp.

supplies, but with one of the inputs connected to 0 V. The above 'op. amp. equation' can be used to find out what the output voltage is in these two cases. If a positive input voltage, V_1, is applied to the noninverting input (pin 3) and the inverting input is connected to 0 V, the output voltage, V_{out}, is positive and equal to $A \times (V_1 - 0)$, i.e. $A \times V_1$. But if a positive input voltage, V_2, is applied to the inverting input (pin 2), the output voltage is given by $A \times (0 - V_2)$, i.e. $-AV_2$. Thus any difference between the two input voltages that makes the noninverting input voltage more positive than the inverting input provides an amplified positive (i.e. above 0 V, and therefore 'noninverted') output voltage. And any difference between the input voltages that makes the inverting input voltage more positive than the noninverting input voltage provides an amplified negative (i.e. below 0 V, and therefore 'inverted') output voltage. It is common to indicate the inverting input of an op. amp. with a '$-$' sign to indicate its ability to invert the sign of an input signal, and the noninverting input with a '$+$' sign to indicate its ability not to invert the sign of an input signal. You should not confuse these signs with the polarities of the power supplies.

Thus we have two basic characteristics of an op. amp.

1 An op. amp has a very high voltage gain, i.e. the 3140 op. amp. used in the thermostat of Fig. 14.2 has a voltage gain in excess of 100 000. Hence op. amps. are sometimes referred to as 'packages of gain'.

2 An op. amp. responds to the difference of voltage between its two input terminals. Hence an op. amp. is sometimes called a difference amplifier.

14.6 Using op. amps as comparators

A comparator is a circuit building block that compares the strength of two signals and provides an output signal when one signal is bigger than the other. Clearly the op. amp. is one type of comparator since its two inputs are able to compare the magnitude of two voltages as shown in Fig. 14.8. It simply compares a voltage, V_{in}, here shown applied to the inverting input of the op. amp., with a reference voltage, V_{ref}, applied to the noninverting input. Thus any slight difference in voltage, $e = V_{ref} - V_{in}$, causes the output voltage to saturate. Since the op. amp. is operated from a dual power supply, the two possible saturation voltages are $+V_{sat}$ or $-V_{sat}$. If a single power supply were used, the saturation voltages would be $+V_{sat}$ and 0 V. To 'saturate' means to go to the maximum value. In

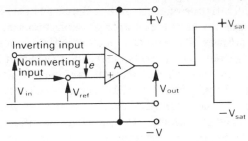

Fig. 14.8 An op. amp used as a comparator

this case, the maximum output voltage is limited to a volt or two below the supply voltage, e.g. 10 V for a 12 V power supply. The output voltage cannot take on any intermediate value between the upper and lower saturation voltages since the gain of the op. amp. is so high. Thus if its gain is 100 000, the difference e between the two input voltages which makes the output voltage saturate at 10 V is (10 V/100 000) or 0.0001 V. If the difference exceeds this value, the output voltage remains saturated at 10 V.

In the design of the thermostat in Fig. 14.2, the reference voltage is applied to pin 3 by two equal-value resistors R_1 and R_2. The voltage divider action of these two resistors sets a reference voltage of about 6 V on the noninverting input, pin 3. This voltage is compared with the changing voltage on the inverting input, pin 2, determined by the resistance of the thermistor. Now suppose the variable resistor VR_1 is set so that the voltage on pin 2 is lower than that on pin 3, say 5.8 V compared with 6 V on pin 3. This makes the output voltage at pin 6 rise to the upper saturation voltage, i.e. about 10 V, and transistor Tr_1 switches on. Thus the relay is energised and power is supplied to the heater. The thermistor senses this rise in temperature and its resistance falls (Section 5.6). The voltage on pin 2 therefore rises. As soon as this voltage exceeds 6 V, the output voltage falls to the lower saturation voltage, 0 V in this case. The transistor switches off and power is no longer supplied to the heater. As the temperature of the thermistor rises and falls, the output voltage of the comparator falls to 0 V and rises to about 10 V, respectively.

But there is one additional component in the design of the thermostat which 'sharpens up' its performance. This is resistor R_3 connected between pin 6, the output, and pin 3, the noninverting input of the op. amp. What does it do? Once the output voltage begins to rise or fall, this change is made all the more rapid by the effect of R_3. Thus the circuit acts rather like a wall light switch, i.e. it snaps off when taken past an upper position, and snaps on when taken past a lower position. The electrical effect of resistor R_3 is shown in Fig. 14.9. The basic comparator is shown in Fig. 14.9a. Two 10 kΩ resistors, R_1 and R_2, set a reference voltage on the noninverting input (point Y) of 6 V. The voltage on the inverting input (point X) is determined by the setting of VR_1 and the temperature of the thermistor, Th_1. Suppose this voltage is set at 6.2 V, which is higher than that at point Y so the output of the op. amp. is LOW and the transistor is switched off. Now if the temperature of the thermistor falls, its resistance rises so that the voltage at point X falls. When it falls below that at point Y, the output voltage of the op. amp. goes HIGH and the transistor is switched on. Note that as the output of the op. amp. goes from LOW to HIGH, there is no change in the reference voltage of 6 V at point Y.

Fig. 14.9b shows the effect of R_3 when the output voltage is LOW. The dotted outline shows that R_3 is now effectively connected in parallel with R_2. This has the effect of lowering the value of R_2, making the reference voltage slightly lower than 6 V. Now in order for the voltage at point X to fall below that at point Y, the temperature of the thermistor must fall slightly lower than it did in the circuit at Fig. 14.9a. When it does, the output voltage of the op. amp. sharply rises to saturation. But note the effect of R_3 on the switching action of the circuit. As soon as the voltage at the output

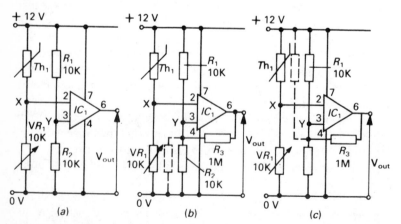

Fig. 14.9 Making a Schmitt trigger from an op. amp. comparator: (a) the basic comparator; (b) the effect of R_3 when $V_{out} = 0$ V; (c) the effect of R_3 when $V_{out} = +V$

starts to rise, the effective value of R_2 increases towards 10 kΩ. This has the effect of raising the voltage at point Y which increases the difference in voltage between points X and Y, thereby making the output move more rapidly towards saturation. This action is called positive feedback, since a proportion of the rising voltage is fed back to the noninverting input of the op. amp where it emphasises the upward trend in the output voltage.

Once the output voltage has risen to +10 V, R_3 is effectively connected in parallel with R_1 as shown by the dotted line in Fig. 14.9c. The voltage at point Y is now slightly higher than 6 V so the temperature of the thermistor must rise slightly higher than it did in the circuit of Fig. 14.9a before the voltage at point X rises above that at point Y. When this happens, the output voltage of the op. amp. falls. And once it starts to fall, the action of R_3 is to make the voltage at point Y fall which increases the difference in voltage between points X and Y thereby making the output voltage fall more rapidly towards 0 V.

This 'snap-action' effect of R_3 in the performance of the circuit gives the circuit the general name of **Schmitt trigger** in honour of the man who first used positive feedback to improve the performance of electrical switching circuits. If R_3 is omitted from the design of the thermostat, the relay contacts are likely to 'chatter' at the on and off switching points. This chatter is due to electrical noise which causes small changes to the voltages at the inputs of the op. amp. and makes the output voltage oscillate. Schmitt triggers are widely used in switching circuits and are available as ready-built circuits in integrated circuit packages.

The difference between the upper and lower values of the voltage at point Y is known as the hysteresis of the Schmitt trigger and is determined by the values of R_1, R_2 and R_3. It is calculated as follows.

When R_3 is in parallel with R_1 ($V_{out} = 0$ V), the effective value of R_1 is 10 kΩ × 1000 kΩ/(10 kΩ + 1000 kΩ) = 9.9 kΩ. Thus the voltage at point X is 12 V × 9.9 kΩ/19.9 kΩ = 5.97 V, i.e. 0.03 V less than the value of 6 V in Fig. 14.9a.

When R_3 is in parallel with R_2 ($V_{out} = 12$ V, say), the effective value of R_2 is 9.9 kΩ as above. Thus the voltage at point X is 12 V × 10 kΩ/19.9 kΩ = 6.03 V, i.e. 0.03 V more than the value of 6 V in Fig. 14.9a. The hysteresis of this Schmitt trigger is therefore 0.06 V.

14.7 The stepping motor

This is a type of electric motor which rotates in precise controlled steps using digital signals. Thus a particular angle of rotation is known precisely from the number of pulses delivered to it. Stepping motors are therefore commonly used in all kinds of computer-controlled equipment, especially robots. Their power outputs range from about 1 kW for industrial applications to a few milliwatts when used to drive the hands of an analogue quartz watch.

The principle of a permanent magnet stepping motor is shown in Fig. 14.10. This has four coil windings, or phases, arranged in pairs, A, $\overline{A}$, B and $\overline{B}$. When a d.c. voltage is applied to phase A, the resultant current sets up a magnetic field in the stator which causes the rotor to align itself with the field as shown by the arrow. If the voltage is then applied to phase B, while phase A is de-energised at the same time, the stator field will shift through 90 degrees and the rotor will move through the same angle to maintain its alignment with the field. Similarly, when phases $\overline{A}$ and $\overline{B}$ are subsequently energised, the rotor moves through two more 90 degree steps to complete one revolution. Of course, a stepping motor with such a big step angle as this is of little use for practical applications.

One common type of low-power stepping motor is the Philips ID35, the basic construction of which is shown in Fig. 14.11. The ID35 is an example of a permanent magnet stepping motor since its rotor is a circular magnet with twelve pairs of magnetic poles round its circumference. And its stator comprises 24 pairs of stator poles

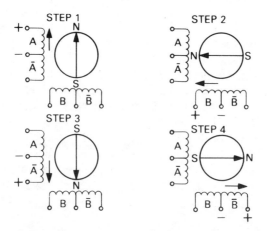

Fig. 14.10 The operating principles of the stepping motor

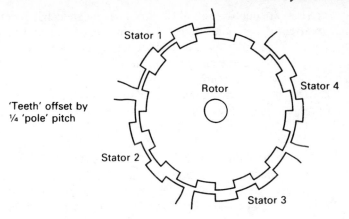

Fig. 14.11 The basic constructional details of the ID35 stepping motor

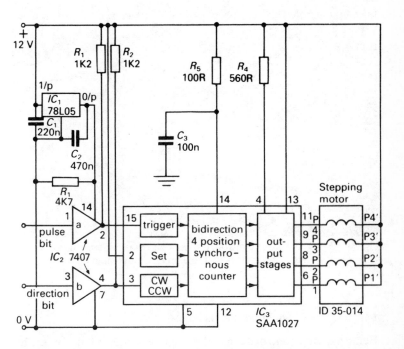

Fig. 14.12 A computer-operated circuit for controlling a stepping motor

arranged as two identical sets of 12 pairs, one set of which is offset by one quarter of the pole pitch. The rotor follows an advancing magnetic field set up when a certain sequence of pulses is fed to the stator field coils. The step angle of the ID35 is 7.5 degrees so it makes 48 steps per revolution.

The ID35 stepping motor can be driven from a computer using a purpose-designed integrated circuit, the SAA1027, as shown in Fig. 14.12. This circuit takes the digital signals from the computer and produces the correct switching sequence on its outputs (pins 6, 8, 9 and 11) to drive the four phases of the stepping motor. A second input, pin 3, selects the direction of rotation of the stepping motor. Thus a HIGH signal on pin 3 causes the motor to rotate one way, a LOW signal the opposite way. Each HIGH/LOW pulse from the computer must be at least 30 ms long and each 7.5° step angle is initiated on the rising edge of the pulse. The circuit operates from a 12 V power supply, though IC_2 requires a 5 V supply which is obtained from the voltage regulator, IC_1. Fig. 14.13 shows a computerised robot which uses powerful stepper motors.

Fig. 14.13   A computer-controlled robot on a car production line that uses stepping motors to make precise movements
Courtesy: Austin Rover

14.8   Digital-to-analogue converters

Being a digital control device, the stepping motor is tailor-made for use with a computer since it is operated by on/off signals that the computer is good at producing. But it may be necessary for the computer to control the speed of analogue motors, or to vary the brightness of a filament lamp, or to synthesise music. These applications often call for the use of a special circuit building block that converts digital signals into equivalent analogue signals. This interface device is called a **digital-to-analogue converter** (DAC) and Fig. 14.14 shows its role as an output device in a computer control system. Note that a DAC is a type of decoder, for it decodes digital information into analogue information. The truth table of a 4-bit DAC is shown below. It gives the analogue equivalent of the sixteen values of the digital data ranging from 0 V for $(0000)_2$, to 2.25 V for $(1111)_2$. Note that a change of 1 bit in the binary information is equivalent to a change of 0.15 V in the output information. Thus if the input information is $(0100)_2$, the output information is $4 \times 0.15 = 0.6$ V and so on.

Digital input (DCBA)	Output voltage (V_{out})
0000	0.00
0001	0.15
0010	0.30
0011	0.45
0100	0.60
0101	0.75
0110	0.90
0111	1.05
1000	1.20
1001	1.35
1010	1.50
1011	1.65
1100	1.80
1101	1.95
1110	2.10
1111	2.25

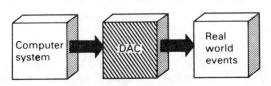

Fig. 14.14 A digital-to-analogue converter is an output device in a computer system

214 *Electronics*

The design of a basic 4-bit DAC is shown in Fig. 14.15. This block diagram has two parts, a resistor network and a summing amplifier based on an op. amp. The resistor network takes into account that a 1 at input B is worth twice as much as a 1 at input A; and a 1 at input C is worth twice as much as a 1 at input B, and so on. Fig. 14.16 shows the type of resistor network required. Each of the resistors is weighted in value in a binary sequence, i.e. R, $2R$, $4R$ and $8R$. The switches would be transistor switches in an actual DAC. They connect the resistors to the 5 V reference voltage if the input bit is a 1 and to 0 V if the input bit is a 0. Suppose the digital input is $(1101)_2$ as shown, since switches 1, 2 and 4 are connected to the 5 V reference voltage. Thus the output voltage from the summing amplifier is given by

$$V_{out} = \frac{V_{ref} \times R_f \left(1 + \frac{1}{2} + \frac{1}{8} \right)}{R}$$

$$= \frac{V_{ref} \times R_f \times 1.625}{R}$$

Since $V_{ref} = 5$ V, $R = 10$ kΩ and $V_{out} = 1.95$ V, $R_f = 2.4$ kΩ approximately.

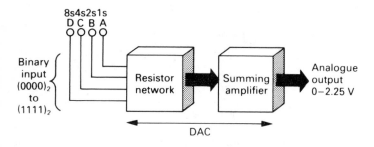

Fig. 14.15   The principle of a simple 4-bit DAC

The problem with this simple DAC is that the resistor network requires a wide ratio of resistor values, i.e. 1 to 128 for the usual 8-bit DAC. If this simple DAC is to perform over the whole range of inputs, the resistors must be close-tolerance types whose values all vary by the same amount with any temperature change they may be subjected to. This problem is overcome by using a **R–2R resistance ladder** as shown in Fig. 14.17a. The technique uses just two resistor values and can be extended to any number of bits. Moreover, the

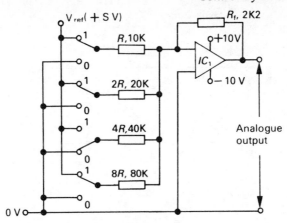

Fig. 14.16 The design of a 4-bit DAC using a binary-weighted resistor network

absolute values of the resistors are unimportant since only their ratio needs to be exactly 2. The truth table below lists the output voltage, V_{out}, as a function of the reference voltage, V_{ref}, for all sixteen possible values of a 4-bit binary number.

Digital input (DCBA)	Output voltage (V_{out})
0000	0
0001	$V_{ref}/16$
0010	$V_{ref}/8$
0011	$3V_{ref}/16$
0100	$V_{ref}/4$
0101	$5V_{ref}/16$
0110	$3V_{ref}/8$
0111	$7V_{ref}/16$
1000	$V_{ref}/2$
1001	$9V_{ref}/16$
1010	$5V_{ref}/8$
1011	$11V_{ref}/16$
1100	$3V_{ref}/4$
1101	$13V_{ref}/16$
1110	$7V_{ref}/8$
1111	$15V_{ref}/16$

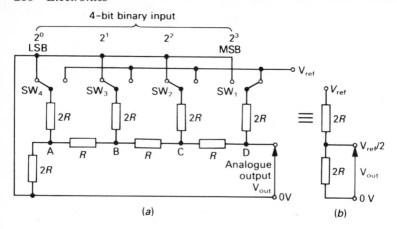

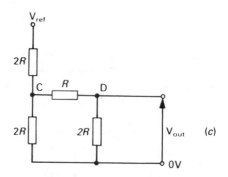

Fig. 14.17 A 4-bit DAC design based on an R–2R ladder: (a) the switch positions with an input of $(1000)_2$; (b) the equivalent resistor network with an input of $(1000)_2$; (c) the equivalent resistor network with an input of $(0100)_2$

To see how these output voltages arise, suppose $(1000)_2$ is input by setting switches SW_1 to logic 1, and SW_2 to SW_4 to logic 0. With this input signal, the entire network to the left of the node D can be replaced by a resistor of value $2R$ so that the equivalent circuit reduces to that of Fig. 14.17b. The output voltage is therefore $V_{ref}/2$ as required since $(1000)_2$ represents half the full-scale input voltage.

Now input the word $(0100)_2$ by setting switches SW_1, SW_3 and SW_4 at logic 0, and SW_2 at logic 1. The network to the left of node C can be replaced by a resistor of value $2R$ as shown in Fig. 14.17c. The voltage at node C is given by $6V_{ref}/16$, and at node D by $V_{ref}/4$

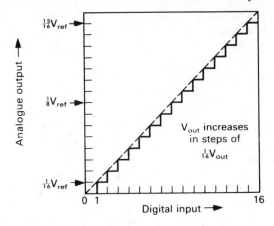

Fig. 14.18 The transfer function of a 4-bit DAC

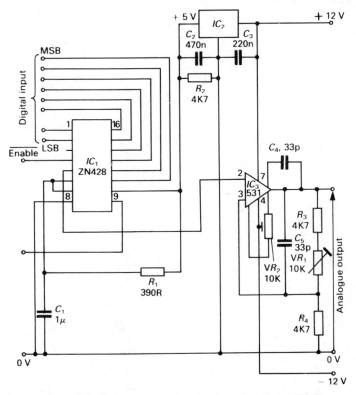

Fig. 14.19 A practical computer-controlled 8-bit DAC

which is the output voltage. This voltage is required since $(0100)_2$ represents a quarter of the full-scale input voltage. Similarly, it is possible to show that a binary input of $(0010)_2$ gives an output voltage of $V_{ref}/8$, and so on.

Fig. 14.18 shows the relationship (known as a **transfer function**) between the digital input voltage and analogue output voltage for this 4-bit DAC. Note there are sixteen discrete values of the output voltage for a 4-bit DAC; an 8-bit DAC has 256 discrete values of output voltage. Thus DACs produce a stepped analogue output voltage. The more bits used, the less coarse the steps for a given output voltage and the smoother the control of the speed of a d.c. motor or of the brightness of a lamp. It is usual to use an integrated circuit package to perform the conversion from digital to analogue voltages. For example, in the Ferranti series of DACs, the ZN426, ZN428 and ZN429 are all 8-bit DACs. Fig. 14.19 is a practical DAC that uses the ZN428 and is suitable for use with any computer that provides an 8-bit data signal.

15

Instrumentation Systems

15.1 Electronics and measurement

Our senses are good at detecting changes in quantities like temperature, frequency and light intensity, but instruments are needed to give us values on which we can all agree. An instrument is not only able to give us the actual value of a quantity, but it can take readings in inaccessible places, such as in the middle of a grain store and on the surface of Mars. Furthermore, instruments can measure quantities such as atomic radiation, radio frequencies and atmospheric pressure, to which our senses are quite oblivious.

Electronic instruments are common. Digital watches, clocks, thermometers and weighing machines are electronic. So, too, are many of the instruments used in weather forecasting and medicine. Car dashboards and the flightdeck of an airliner contain many different types of electronic instrument. The systems diagram shown in Fig. 15.1 summarises the general function of these instruments. (See also Section 4.6.) An instrumentation system comprises three basic building blocks: sensor, signal processor and display. It may have an analogue or a digital display. This model summarises the function of the two instruments described in this chapter: a thermometer for temperature measurement and a Geiger counter

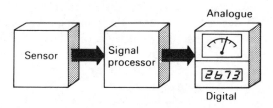

Fig. 15.1 The three main building blocks of an instrumentation system

for measuring radioactivity. First let's look at the characteristics of two sensors suitable for use in these instruments.

15.2   Thermocouples and Geiger tubes

(a)   *Thermocouples*

For accurate temperature measurement over a wide temperature range, thermocouples are generally preferred to semiconductor devices such as diodes (Chapter 7) and thermistors (Chapter 5). A thermocouple is made from two dissimilar metals or alloys, A and B, joined together to form two junctions, J_1 and J_2 as shown in Fig. 15.2a. An e.m.f. is generated when there is a difference of temperature between these two junctions. Fig. 15.2b shows that the variation of e.m.f. with temperature is generally linear over a wide temperature range. Commercial thermocouples are available for measuring temperature in the range −200 °C to 1000 °C. These thermocouples are classed as type T and type K. Type T thermocouples are based on the combination of copper and copper/nickel and operate in the range 0 °C to 1100 °C. Type K thermocouples are based on the combination of the alloys nickel/chromium and nickel/aluminium (like the one shown in Fig. 15.2b) and are used in the range −200 °C to +400 °C.

(b)   *Geiger tubes*

Radioactive materials such as plutonium and uranium emit three main types of nuclear radiation: alpha, beta and gamma. Alpha and beta radiation are made up of particles, alpha radiation comprising the nuclei of helium atoms (two protons and two neutrons), and beta radiation comprising electrons (see Section 2.2). Gamma radiation is a high frequency electromagnetic wave (see Fig. 16.2). Alpha particles have the least penetrating power, being readily stopped by paper and skin. However, alpha radiation causes damage if it comes from a radioactive substance that has somehow got into the body, for instance by breathing it in or eating contaminated food. Beta particles lose their energy within several metres in air and can easily be stopped by a few millimetres of aluminium. Gamma radiation has the most penetrating power but it too can be stopped by thick concrete or lead.

There are many types of sensor for detecting these different types of nuclear radiation. The one described here is the Geiger tube (strictly a Geiger–Müller tube) and is named after Hans Geiger who

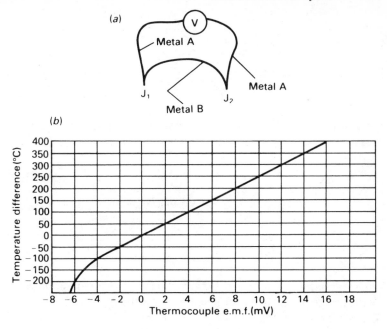

Fig. 15.2 (a) The principle of a thermocouple, and (b) the e.m.f. produced by a type K thermocouple

lived from 1882 to 1947. What a Geiger tube responds to and counts is the passage of an alpha or beta particle, or a gamma ray, through it. A Geiger tube is very simple in principle as shown in Fig. 15.3. It consists of an anode wire held at a high positive potential compared with a cathode shield that surrounds the wire. The anode and cathode are usually enclosed in a glass envelope which is filled with an inert gas such as krypton. A resistor is used in series with the anode.

When an incoming particle of radiation passes through the tube, the gas in it is briefly ionised. This reduces the resistance between the anode and the cathode and the anode potential falls. One of the gases contained in the tube is a 'quenching agent' to ensure that once the particle has passed through the tube, the potential at the anode end of the tube rises sharply to its normal value. The brief fall in potential across the tube is then amplified and fed to a monitoring circuit. Typically the anode potential needed for the tube to work is between 400 V and 1000 V. But precise voltage control is unnecessary in a simple Geiger counter since an individual Geiger tube has a

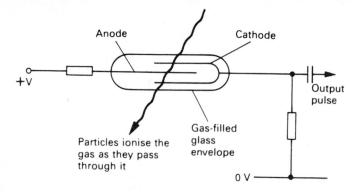

Fig. 15.3 The structure of a Geiger tube

wide operating voltage range as shown in Fig. 15.4. Below about 400 V the anode potential is insufficient to allow the gas to ionise. At the Geiger threshold, the tube begins to 'count' and continues to do so over a wide range of anode potential called the plateau. As the operating potential increases, the sensitivity of the tube increases slightly. It is important not to increase the anode potential beyond the plateau since this will shorten the life of the tube considerably. A Geiger tube operated correctly remains useful for at least 500 billion counts.

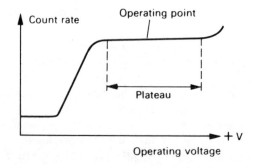

Fig. 15.4 How the sensitivity of a Geiger tube varies with its operating voltage

15.3 The design of an electronic thermometer

Three building blocks form the basis of the electronic thermometer shown in Fig. 15.5. The thermocouple senses the temperature and converts temperature into a small e.m.f. between its 'cold' and 'hot' junctions. This e.m.f. is only a few millivolts so a voltage amplifier, IC_1, is used to enable a 1 V d.c. voltmeter, V_1, to read the temperature. IC_1 is an integrated circuit called an operational amplifier, the characteristics of which were explained in Section 14.5. Junction J_1 of the thermocouple senses temperature relative to a 'cold' junction where the two wires A and B join the circuit. Clearly, in this simple circuit the ambient temperature of the circuit must not alter between measurements.

The thermometer is calibrated by first deciding the upper and lower readings to be taken. Suppose these are 0 °C, the melting point of ice, and 100 °C, the boiling point of water at sea level. First, junction J_1 is placed in crushed melting ice and the preset potentiometer, VR_1, is adjusted so that the meter reads 0 V corresponding to 0 °C. Next, junction J_1 is placed in boiling water and VR_2 adjusted so that the meter reads 1 V. This second variable resistor alters the voltage gain of the amplifier as explained in Section 15.5. You would need to check these two readings a couple more times, since the adjustments are not completely independent of each other. Once the upper and lower readings are set, the scale of the

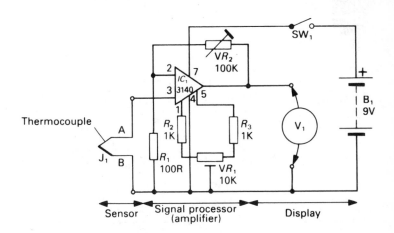

Fig. 15.5 The circuit design of an electronic thermometer

voltmeter can be calibrated in degrees Celsius, i.e. half scale
reading (0.5 V) corresponds to 50 °C. A 1 V d.c. digital voltmeter
gives the actual temperatures (ignoring the decimal point), i.e. 0.18
V equals 18 °C. Here we are assuming that the output of the
thermocouple is proportional to temperature. This is a fair assump-
tion as the graph of Fig. 15.2b shows.

15.4 The design of a Geiger counter

Three building blocks are used in the simple design of the Geiger
counter shown in Fig. 15.6. The first building block converts the 9 V
d.c. battery e.m.f. into an adjustable d.c. voltage of between 450 V
and 550 V to operate the Geiger tube – this circuit is called a
d.c.-to-d.c. converter. It comprises an oscillator based on a 555
timer, IC_1, wired up as an astable (Section 6.7) which feeds a
varying voltage to the secondary winding of a low voltage mains
transformer, T_1. A varying a.c. voltage of approximately 250 V
appears across the primary winding of this transformer. This voltage
is rectified and doubled using diodes D_1 and D_2 and capacitors C_3

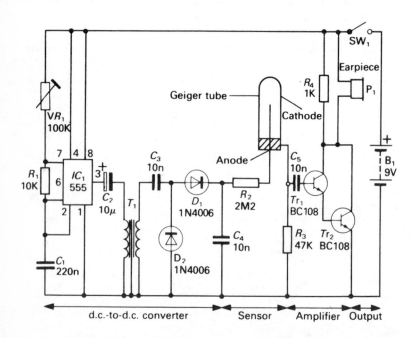

Fig. 15.6 The circuit design of a Geiger counter

and C_4 to provide an adjustable d.c. voltage for operating the Geiger tube.

This d.c. voltage is applied to the Geiger tube via the series resistor R_2. This resistor is essential for correct operation of the Geiger tube. It is usual to operate the Geiger tube about half-way along its plateau (Fig. 15.4). This is done by adjusting VR_1 in the astable circuit which has the effect of 'tuning' the transformer and making it more or less efficient – a rather crude method of achieving the correct operating voltage but it works well enough in this simple design. A high resistance voltmeter must be used to measure the voltage applied to the anode of the tube. Note that this circuit design reflects much more complex commercial designs: a building block to provide a high d.c. voltage for operating the Geiger tube; a pulse amplifier circuit; and a recording device, in this case a crystal earphone, P_1.

Building block 3 is a simple low-power amplifier based on transistors Tr_1 and Tr_2 wired up as a Darlington pair (Section 8.8). Nuclear radiation that passes through the tube ionises the gas in it and causes its resistance to fall. This produces a sharp rise in voltage across resistor R_3 followed by a sharp fall in this voltage as the ionisation in the tube is rapidly quenched. The voltage pulse produced across R_3 is coupled to the Darlington pair via capacitor C_5. Thus each time an alpha or beta particle, or a 'quantum' of gamma radiation, passes through the Geiger tube, a sharp click is heard in the crystal earphone, P_1. Random clicks are heard even when there is no man-made radiation source nearby. This 'background radiation' comes from radioactive gases in the air, from radioactive materials in earth and buildings, and from cosmic rays which enter the Earth's atmosphere with great energy and produce secondary 'showers' of nuclear particles that reach the ground. The leakage of radiation from a damaged nuclear power station, such as the one at Chernobyl in the USSR, and from the disposal of radioactive wastes underground and at sea may contribute to this background radiation.

15.5 Designing voltage amplifiers with op. amps

The basic properties of op. amps and their use as comparators and Schmitt triggers in control systems have been described in Chapter 14. Op. amps are also widely used in the design of instruments. For example, the electronic thermometer described in Section 15.3 uses an op. amp. to amplify the small e.m.f. generated by a thermocouple temperature sensor. In this application, the intrinsically high voltage gain of an op. amp., which is so useful in the design of control systems, has been 'tamed' to provide a much reduced but accurately known voltage gain. How is this possible?

The technique used to control the voltage gain of an op. amp. is called **negative feedback**. (You will remember that positive feedback is used in the design of a Schmitt trigger – see Section 14.6.) There are two basic voltage amplifier circuits that make use of negative feedback: the inverting negative feedback voltage amplifier (Fig. 15.7a); and the noninverting negative feedback voltage amplifier (Fig. 15.7b). In these circuits the ratio of the output voltage V_{out} to the input voltage V_{in} is the voltage gain A of the amplifier. The equations which determine this voltage gain are remarkably simple as the following table shows.

Amplifier circuit	Voltage gain, $A =$
inverting negative feedback voltage amplifier	$\dfrac{V_{out}}{V_{in}} = \dfrac{-R_2}{R_1}$
noninverting negative feedback voltage amplifier	$\dfrac{V_{out}}{V_{in}} = 1 + \dfrac{R_2}{R_1}$

Note two things about these equations:

1 First, neither equation makes reference to the intrinsically high gain of the op. amp.; the gain in each case is determined only by the values of the external resistors – a surprising result. You might be tempted to suggest removing the op. amp. from the circuits leaving only the two resistors!
2 Second, the negative sign in the equation for the inverting voltage amplifier means that the input voltage is inverted, i.e. a positive input voltage produces a negative output voltage, and vice versa. The noninverting voltage amplifier does not change the sign of the input voltage.

Thus supposing you choose resistor values of $R_2 = 100$ kΩ and $R_1 = 10$ kΩ. In the case of the inverting amplifier the voltage gain is -100 k$\Omega/10$ k$\Omega = -10$ times. And in the case of the noninverting amplifier, the voltage gain is $1 + 100$ k$\Omega/10$ k$\Omega = 11$ times. An input voltage of $+1$ V produces an output voltage of -10 V in the inverting amplifier, and of $+11$ V in the noninverting amplifier.

(Here we are assuming that the op. amps are operated from a dual power supply so that the output voltage is negative, i.e. below 0 V.) Of course any other values of resistors R_2 and R_1 could be used to obtain the voltage gain required for a particular application. Note that the thermocouple thermometer shown in Fig. 15.5 is connected as a noninverting amplifier. The e.m.f. generated by the thermocouple is fed to the noninverting input, pin 3, of the op. amp. It is amplified by a varying factor depending on the setting of VR_1. Thus if VR_1 is set to its maximum value, the voltage gain is given by

$$1 + VR_1/R_1 = 1 + 100 \text{ k}\Omega/0.1 \text{ k}\Omega = 1001 \text{ times.}$$

So if at a particular temperature the e.m.f. generated by the thermocouple is about 2 mV (2/1000) V, the output voltage recorded on the meter is about 2 V.

But we still haven't explained how the circuits shown in Fig. 15.7 manage to control the intrinsically high gain of the op. amp., i.e. from values of 100 000 or more (known as the open-loop voltage gain, A_{vol}) to less than 1000, say (known as the closed-loop voltage

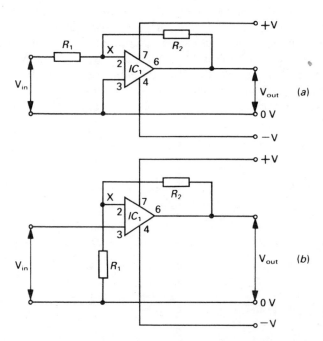

Fig. 15.7 The use of op. amps as negative feedback voltage amplifiers: (a) an inverting amplifier, and (b) a noninverting amplifier

gain, A_{vcl}). To explain why, we need to assume that op. amps have two ideal characteristics. These are:

1 they have an infinitely large open-loop voltage gain, i.e. $A_{vol} \rightarrow$ infinity (actual op. amps have open-loop gains in excess of 100 000).
2 they draw no current whatsoever from the source of the signals at either of their two inputs, i.e. $I \rightarrow$ zero (actual op. amps have input currents less than 10^{-6}A).

These ideal characteristics enable us to prove the two closed-loop gain equations. First, let's start with the inverting voltage amplifier shown in Fig. 15.7a in which the input voltage, V_{in}, is applied to the inverting input of the op. amp. via resistor R_1. Resistor R_2 is a 'feedback resistor' which 'closes the loop' between the output (pin 6) and the inverting input (pin 2). The input side of R_1 is at a voltage of V_{in}, and the output side of R_2 is at a voltage of V_{out}, both voltages measured with respect to 0 V. So what is the voltage, V_X, at the join between the two resistors, i.e. at point X, the inverting input of the op. amp.? Fig. 15.8a shows the connections at the ends of the resistors.

Note that the noninverting input of the op. amp. is connected to 0 V. Now if we have a perfect op. amp. (i.e. one that has an infinitely high open-loop gain), there is no difference between the two input voltages: i.e. if pin 3 is at 0 V, pin 2 must be at 0 V. Of course, for practical amplifiers like the 741 and 3140, a very small difference of voltage (less than a microvolt) does exist between the two inputs. Thus for the perfect op. amp. the voltage V_X at point X in Fig. 15.8a

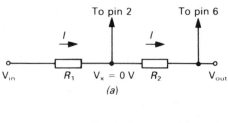

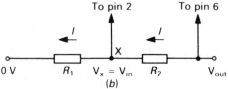

Fig. 15.8   Proving the closed-loop gain equations for (a) the inverting amplifier, and (b) the noninverting amplifier

is 0 V. This point isn't actually connected to 0 V but it might just as well be – and it is therefore called the **virtual earth** in the op. amp. circuit.

If we assume our op. amp. does not require any input current, we can concentrate our attention on working out the relationship between V_{out} and V_{in}. Let's assume that V_{in} is positive so that a current I flows through R_1 towards X. This same current flows through R_2 which is in series with it, since no current flows into pin 2. Thus Ohm's law can be used to write down the following equations.

For resistor R_1: $I = (V_{in} - 0)/R_1$
For resistor R_2: $I = (0 - V_{out})/R_2$

Note that the voltage difference across R_2 is $0 - V_{out}$ since current flows from pin 2 to pin 6. These equations give $(V_{in}/R_1) = -V_{out}/R_2$ which can be rearranged to give

$$(V_{out}/V_{in}) = -(R_2/R_1) = A_{vcl},$$

the equation which was written down above.

Now let's look at the noninverting amplifier shown in Fig. 15.7b. In this case, the input voltage V_{in} is applied direct to the noninverting input, pin 3, and R_1 is connected from pin 2 to 0 V. Note that R_1 and R_2 are connected in series as shown in Fig. 15.8b. The voltage at one end of R_2 is V_{out}, and at one end of R_1 is 0 V. Again, before we can use Ohm's law to prove the gain equation, we have to know the voltage, V_X, at point X between the two resistors. Since the op. amp. is perfect, there is no difference of voltage between the two inputs. So as V_{in} changes, V_X must follow these changes, i.e. $V_X = V_{in}$. Note that in this case neither pin 2 nor pin 3 is at 0 V, i.e. virtual earth. Let's assume that a current I flows through R_1. This same current (remember there is no current into pin 2) then flows through R_2 which is in series with it. Thus the following equations can be written down.

For resistor R_1: $I = (V_{in} - 0)/R_1$
For resistor R_2: $I = (V_{out} - V_{in})/R_2$

These equations give $V_{in}/R_1 = (V_{out} - V_{in})/R_2$. Dividing through by V_{in} and rearranging gives

$$V_{out}/V_{in} = 1 + R_2/R_1 = A_{vcl},$$

the equation which was written down above.

Thus the closed-loop voltage gain of negative feedback amplifiers is independent of variations in the characteristics of the resistors, transistors and other discrete components which make up the circuit inside the op. amp. It doesn't matter whether we use an op. amp. with a gain of 100 thousand, or 1 million, or 250 thousand, the closed-loop voltage is determined solely by the values of resistors connected externally to the op. amp. The feeding back of part of the output voltage to the inverting input is known as negative feedback and its effect is to stabilise the voltage gain of the op. amp. to a value determined by the values of the external resistors. Thus if the output voltage tends to rise, a small fraction of this rise is applied to pin 2. Since this is the inverting input of the op. amp., the op. amp. tries to lower the output voltage. Likewise, a decrease in the output voltage is counterbalanced by the op. amp. trying to raise the output voltage. The only stable value of the output voltage is that determined by the values of the two external resistors.

There are other advantages of using negative feedback in amplifiers like the op. amp. One is that it increases the bandwidth of the amplifier. As explained in Section 8.3, the bandwidth of an amplifier is the range of frequencies within which its voltage gain is not less than 0.7 of its maximum value. The curves in Fig. 15.9 are for the 741 op. amp., and show the voltage gain in decibels. The open-loop voltage gain of the 741 is about 100 dB. Now 100 dB is equal to $20\log_{10}(A_{vol})$, i.e. $A_{vol} = 100\,000$. When no feedback is applied, i.e. the op. amp. is operated 'open loop' and $A_{vol} = 100$ dB, the upper curve in Fig. 15.9 shows that the bandwidth extends

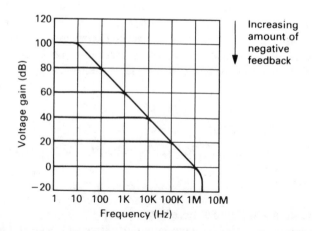

Fig. 15.9   The gain versus bandwidth curves for the 741 op. amp.

from 0 Hz, i.e. d.c., to just a few Hz. Thus the op. amp. would be quite useless as an audio amplifier if we wanted to use all the voltage gain it's capable of giving. But if we design a negative feedback voltage amplifier which has a closed-loop voltage gain of, say, 40 dB (i.e. 100 times), the bandwidth available is now about 10 kHz and the op. amp. would give quite a respectable performance as an audio amplifier.

15.6 Analogue-to-digital converters

In Section 14.8 it was explained that a digital-to-analogue converter (DAC) is a type of decoder used at the output of digital systems such as computers. It converts digital information into analogue information, thereby enabling digital systems to 'talk' to the real world. As shown in Fig. 15.10, an analogue-to-digital converter (ADC) is used at the input of digital systems; it is a type of encoder which converts analogue information into digital information thereby enabling the digital system to 'listen in' to the real world. Thus, before a computer can 'understand' analogue quantities such as temperature, pressure and wind speed, an ADC is required. ADCs are used in all types of digital instrumentation, such as digital thermometers and multimeters.

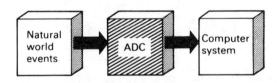

Fig. 15.10 An ADC is an encoding device

Most ADCs comprise two main building blocks, a comparator and digital logic circuits as shown in Fig. 15.11. The truth table below shows that this simple design converts an analogue voltage ranging, say, from 0 V to 2.4 V into a 4-bit digital output ranging from $(0000)_2$ to $(1111)_2$. Thus an input voltage of 1.8 V is converted into a digital output of $(1100)_2$, and so on.

Analogue input (V_{in})	Digital output (DCBA)
0.00	0000
0.15	0001
0.30	0010
0.45	0011
0.60	0100
0.75	0101
0.90	0110
1.05	0111
1.20	1000
1.35	1001
1.50	1010
1.65	1011
1.80	1100
1.95	1101
2.10	1110
2.25	1111

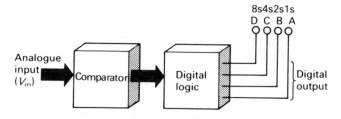

Fig. 15.11 The principle of a 4-bit ADC

The principle of one type of ADC is shown in Fig. 15.12a. It is called a **single-ramp and counter ADC**. The waveforms in Fig. 15.12b show that the ramp generator produces an output voltage rising uniformly and then falling sharply before repeating. This ramp signal is applied to the noninverting input of the comparator where it is compared with the steady analogue voltage on the inverting input of the comparator. While the ramp voltage is less than the input voltage, a counter counts pulses generated by the clock generator. As soon as the voltage of the ramp signal equals the input voltage, the counter stops counting and the number of pulses it has accumulated is proportional to the input voltage.

The operating sequence is shown by the waveforms. First, the

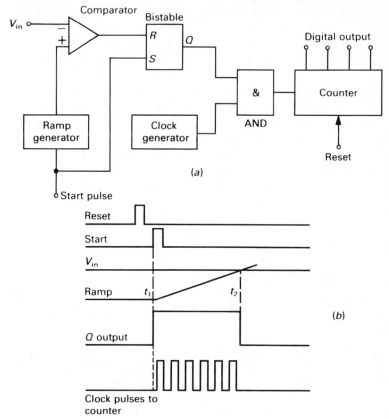

Fig. 15.12 (a) The block diagram and (b) waveforms of a single-ramp and counter ADC

counter is set to zero using a reset pulse. Then at time t_1 a 'start' pulse makes the ramp voltage rise and at the same time sets the Q output of the bistable HIGH, thereby opening the AND gate and allowing pulses from the generator to reach the counter. Clock pulses are accumulated by the counter until the ramp voltage equals the input voltage, V_{in}. At a time t_2, the output of the comparator rises HIGH and resets the Q output of the bistable LOW. This signal closes the AND gate so that counts are prevented from reaching the counter. The digital output from the counter is then proportional to the analogue input voltage. This digital output could be fed to a computer for processing, or converted into a binary-coded decimal (BCD) format to operate a digital display as in a digital voltmeter.

The single-ramp and counter suffers from two drawbacks: it is slow to convert analogue voltages to digital signals, and it needs a highly stable clock generator. For faster and more stable operation, it is usual to use a **dual-ramp and counter ADC** the principle of which is shown in Fig. 15.13a. It does not need a highly stable clock generator, and a building block called an integrator is used instead of a comparator. The waveforms in Fig. 15.13b show its operating sequence. At the start, switches SW$_1$ and SW$_2$ are open and the counter is reset to zero. The control logic then operates SW$_1$ so that the input voltage, V_{in}, is fed to the integrator. This generates a negative-going ramp which has a slope of $-V_{in}/RC$. Another building block called a zero-crossing detector sends a HIGH to the AND

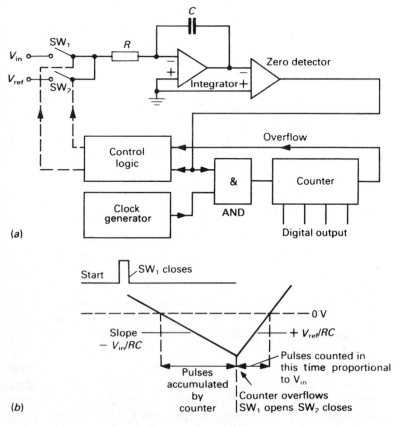

Fig. 15.13 (a) The block diagram and (b) waveforms of a dual-ramp and counter ADC

gate when it has detected that the ramp signal has passed through zero. The AND gate now allows clock pulses to pass to the counter.

The counter accumulates clock pulses until it reaches a maximum count of 2^n where n is the number of bits being converted by the ADC, i.e. a count of 256 for an 8-bit ADC. As soon as the counter overflows and returns to zero, it sends a signal to switch off SW_1 and switch on SW_2 so that the reference voltage, V_{ref}, is now applied to the integrator. The output of the integrator now generates a positive-going ramp which starts from the previous negative value (which was proportional to V_{in}) with a slope of $+V_{ref}/RC$. The counter begins to count until the integrator output again crosses zero. At this point, the zero-crossing detector switches LOW and closes the AND gate. The time taken for the positive-going ramp to reach zero from the previous negative value is proportional to V_{in}. Hence the number of counts accumulated in this time is proportional to V_{in}. The cycle then repeats, and a fresh conversion of the input voltage is produced.

One of the main requirements of an ADC is that it should produce a digital equivalent of an analogue voltage as quickly as possible. ADCs based on the ramp-and-counter process are inherently slow since the counter takes time to accumulate the count. The **successive-approximation counter** shown in Fig. 15.14a is faster. It comprises a voltage comparator, a digital-to-analogue converter (DAC), a logic programmer, and a register. At the start of the conversion, the most significant bit (MSB) is applied to the DAC. The output of the DAC is compared with the analogue input voltage, V_{in}. The MSB is left in or taken out depending on the output of the comparator. If the DAC output is larger than V_{in}, the MSB is removed and placed in the next most significant bit for comparison. The process is repeated down to the least significant bit (LSB) and at this time the required number is in the counter.

Thus the successive-approximation method is the process of trying one bit at a time, beginning with the MSB as shown on the flow chart of Fig. 15.14b for a 4-bit conversion of an analogue input voltage. Suppose this voltage is 7 V $(0111)_2$. First, the MSB is set to 1 (block 1) and the logic circuit feeds the binary number $(1000)_2$ to the DAC. The DAC sends the analogue equivalent of this binary number to the comparator which answers the question 'Is (1000) too high or too low?' (block 2). The answer 'Too high' is sent to the logic circuit (block 3) which cancels the MSB and sets the next MSB to 1 so that the number $(0100)_2$ is sent to the DAC and its analogue equivalent to the comparator. The answer to the question 'Is

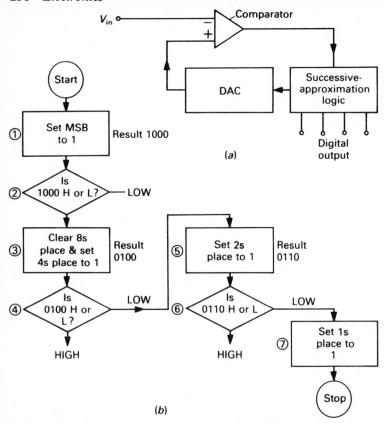

Fig. 15.14 (a) The block diagram and (b) sample flowchart of a successive-approximation ADC

$(0100)_2$ too high or too low?' (block 4) is 'Too low'. And this is sent to the logic circuit which records the second most significant bit as a 1, and then sets the third most significant bit to a 1 (block 5). The comparison of this number with V_{in} yields the answer 'Too low', and the third bit is set to a 1 so that $(0110)_2$ is now stored. Finally a 'guess' of a 1 for the last bit sets the last bit to a 1 and finally yields the number $(0111)_2$.

A practical ADC is shown in Fig. 15.15 and is based on IC_1 which is a purpose-designed ADC. The potentiometer VR_1 sets the zero of the ADC and the variable resistor VR_2 the gain. Thus this circuit can be set to convert analogue voltages between 0 V and 2.55 V into 256 equivalent-value 8-bit words from $(00000000)_2$ to $(11111111)_2$.

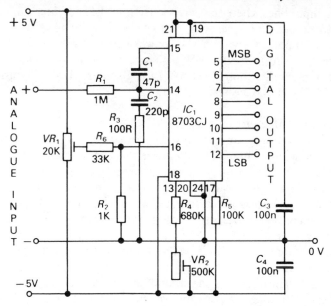

Fig. 15.15 A practical 8-bit ADC based on a purpose-designed integrated circuit

It will therefore accept analogue voltages from the thermocouple thermometer described in Section 15.3 and these can be fed into a microcomputer that does not have a built-in ADC.

15.7 Computer programs

In this chapter the designs for a thermocouple thermometer and Geiger counter have been described. The thermometer provides a continuous or analogue signal and the Geiger counter an on/off or digital signal. Most microcomputers can be used to process these signals so that graphical displays of temperature or count rate can be obtained on the computer's VDU. The programs below are written in BASIC for the BBC microcomputer. The voltage from the thermometer is input via a connector unit which plugs into the BBC's ADC via its analogue port situated at the rear of the machine. The pulses from the Geiger counter are fed into the user port of the BBC using a similar connector unit. The thermometer programs use channel 0 (CH0) of the BBC microcomputer's four-channel ADC. The Geiger counter program uses bit 2 of the 8-bit data input lines where they are loaded into the computer's internal 16-bit counter for processing. Both circuits can be powered by this microcomputer's own 5 V d.c. power supply. Suitable connector units in kit form for the BBC microcomputer are available from Technology Teaching Systems Ltd, Penmore House, Hasland Road, Hasland, Chesterfield, S41 0SJ.

Program 1 Max/min thermometer
This simple program displays the current temperature and the maximum and minimum temperature reached in a period of time.

Lines 30 and 60 display the three temperatures.

Line 70 assigns a variable to the output voltage from the thermometer, which is accessed via channel 1 of the analogue port. Division by 360 converts the analogue reading to a number equal to temperature.

Line 100 prints the initial value of *A* at all three positions of the temperature display.

Line 110 introduces a one second time delay between readings.

Line 120 takes a new value of the input voltage representing temperature.

Line 130 decides whether or not the temperature has changed; if it has, it directs the program to a procedure at line 140 to update the maximum and minimum temperatures, otherwise the program loops back to line 100 and continues to display an unchanged result. The program continues indefinitely between lines 90 and 180.

```
10   REM "MAX/MIN THERMOMETER"
20   CLS
30   PRINTTAB(1,6);"MAX/MIN THERMOMETER"
40   PRINTTAB(4,12); "NOW TEMP=      C"
50   PRINTTAB(4,14); "MAX TEMP=      C"
60   PRINTTAB(4,16); "MIN TEMP=      C"
70   A=INT(ADVAL(1)/360)
80   P=A:R=A:Q=A
90   REPEAT
100  PRINTTAB(13,12);P;TAB(13,14);Q;TAB(13,16);R
110  I=INKEY(100)
120  X=INT(ADVAL(1)/360)
130  IF X<>P THEN PROCmaxmin ELSE 100
140  DEFPROCmaxmin
150  IF X>Q THEN Q=X ELSE Q=Q
160  IF X<R THEN R=X ELSE R=R
170  P=X
180  UNTIL FALSE
```

Program 2 Temp/time plot

This program produces a graph of temperature against time.

Line 10 selects a graphics mode that displays large-size printing on the VDU.

Lines 20, 30 and 40 identify three procedures that structure the program.

The procedure at line 50 designs a graphics window for the graph and prints labels for the graph's axes.

The procedure at line 110 draws in the axes of the graph.

The procedure at line 150 plots the graph after the user has decided (lines 160, 170 and 250) what interval of time is needed between readings.

Lines 200 and 210 select the reading from channel 1 of the analogue port and convert it to a number equal to temperature.

Line 220 prints the value of the temperature in the graphics window.

Lines 230 and 240 plot points on the graph.

When plotting is complete, line 260 selects a procedure that asks whether readings are to be repeated.

```
  1   REM "TEMP/TIME PLOT"
 10   MODE5
 20   PROCwindow
 30   PROCaxes
 40   PROCplot
 50   DEFPROCwindow
 60   CLS
 70   VDU 24,30;400;1179;1000;:GCOL 0, 130
 80   CLG
 90   VDU5:GCOL0,0:MOVE55,940:PRINT"t":MOVE55,880:
      PRINT"e":MOVE55,820:PRINT"m":MOVE 55, 760:PRINT"p":
      MOVE860,450:PRINT"time":GCOL0,1:MOVE350,950:
      PRINT"TEMP/TIME":VDU4
100   ENDPROC
110   DEFPROCaxes
120   GCOL0,0:MOVE 120, 460:DRAW 1080,460:MOVE 120,
      460:DRAW 120, 960
130   FORX=240 TO 1080 STEP 120:MOVE
      X,460:DRAWX,470:NEXTX
140   ENDPROC
150   DEFPROCplot
160   PRINTTAB(0,25);"PLOT RATE?:ENTER SEC"
170   INPUT S
180   F=130
```

```
190   REPEAT
200   CH0=ADVAL(1)
210   N=CH0/360
220   PRINTTAB(9,4);INT(N)
230   Y=460+5*N
240   PLOT 69,F,Y
250   I=INKEY(100*S)
260   IF F>1080 THEN PROCagain
270   F=F+10
280   UNTIL FALSE
290   DEFPROCagain
300   PRINTTAB(0,25);"PLOT AGAIN?YES OR NO"
310   A$=GET$
320   IF A$="Y" THEN 20
330   IF A$="N" THEN PRINTTAB(0,25);"PLOTTING IS FINISHED"
```

Program 3 Geiger counter

Five procedures at lines 90, 160, 270, 330, 430, 510, 550, 680 and 760 control this program, which plots a ten-stage bar graph indicating the counts accumulated in a chosen period of time, and prints the count rate and the average count.

The selection of sampling rate and vertical scaling of the graph is achieved by lines 270 to 320.

Lines 330 to 420 print numbers 1 to 9 along the bottom of the screen at the graphics cursor for more accurate positioning of these numbers.

Lines 430 to 500 define the window at the top of the screen in which data is printed.

Lines 510 to 540 print the present count rate and the average count rate in the graphics window. Note that after each sample, numbers only are printed in the gaps to increase computing speed and to prevent flicker.

Lines 550 to 670 count the pulses received via bit 2 of the input port. Lines 580 and 590 set the BBC's internal counter to 65535. Line 600 provides the chosen delay between sampling. Line 610 then finds the number left in the internal counter which holds a 16-bit number. Line 620 works out the number of counts. Lines 680 to 750 plot the result.

Line 640 updates the total count. Lines 770 to 790 print the results in the graphics window.

Lines 90 to 150 ask the user whether he or she wants the sampling to be repeated.

```
10    REM "GEIGER COUNTER"
20    PROCset-up
30    MODE5
40    PROCstart-screen
50    MODE1
60    PROCbar-numbers
70    PROCwindows
80    PROCprint-results-mask
90    PROCcount
100   VDU7
110   PRINTTAB(1,3);"Do You Want To Repeat This?'Y' or 'N'"
120   A$=GET$
130   IF A$="Y"GOTO20
140   IF A$="N"THEN PRINTTAB(1,3);"Run this program again if
      you want to"
150   END
160   DEFPROCset-up
170   total%=0
180   count%=65128
190   acr%=65131
200   pcr%=65132
210   ddr%=65122
220   ?ddr%=0
230   ?acr%=32
240   ?pcr%=0
250   @%=1
260   ENDPROC
270   DEFPROCstart-screen
280   PRINTTAB(3,2);"GEIGER COUNTER";TAB(2,7);"HOW MANY
      SECONDS";TAB(2,9);"BETWEEN SAMPLES?"
290   INPUThowlong
300   PRINTTAB(1,15);"WHAT SCALE FACTOR?"
310   INPUTscale%
320   ENDPROC
330   DEFPROCbar-numbers
340   VDU5
350   MOVE17,45
360   FOR N=1 TO 9:PRINTN;
370   PLOT0,105,0
380   NEXTN
390   PLOT0,-25,0
400   PRINT10
410   VDU4
420   ENDPROC
430   DEFPROCwindows
440   VDU24,0;85;1279;850;
```

```
450   VDU28,0,5,39,1
460   COLOUR129
470   CLS
480   @%=&20205
490   MOVE0,85
500   ENDPROC
510   DEFPROCprint-results-mask
520   PRINTTAB(1,0);"Present Count Rate=   ";"counts/sec"
530   PRINTTAB(1,1); "Average Count Rate=   ";"counts/sec"
540   ENDPROC
550   DEFPROCcount
560   FORsample%=1 TO 10
570   TIME=0
580   ?count%=255
590   ?(count%+1)=255
600   REPEAT:UNTIL TIME>howlong*100
610   result%=?(count%+1)*256+?count%
620   result%=65535-result%
630   PROCbar(result%*scale%)
640   total%=total%+result%
650   PROCprint-results
660   NEXTsample%
670   ENDPROC
680   DEFPROCbar (high%)
690   PLOT80,0,high%
700   PLOT80,60,0
710   PLOT81,0,-high%
720   PLOT80,-60,0
730   PLOT81,0,high%
740   PLOT80,136,-high%
750   ENDPROC
760   DEFPROCprint-results
770   PRINTTAB(22,0)result%/howlong
780   PRINTTAB(22,1)total%/sample%howlong
790   ENDPROC
```

16

Telecommunications Systems

16.1 Introduction

Telecommunications began when the electric telegraph was invented at the beginning of the 19th century. It is now the world's fastest growing industry and includes radio, television and optical communications. All telecommunications systems have one thing in common: the messages they send are converted into signals that can be transmitted through wires, interplanetary space and even glass fibres. Thus 'telecommunications' simply means 'communication at a distance'.

Fig. 16.1 shows the basic building blocks of a telecommunications system. It comprises a communications channel along which a message is transferred between a transmitter and a receiver. In the transmitter, a transducer (e.g. a microphone) takes the message from a source and puts it into a suitable form for transmission along the communications channel. In the receiver, another transducer (e.g. an earphone) delivers a copy of the transmitted information to a destination. Other building blocks might be part of this com-

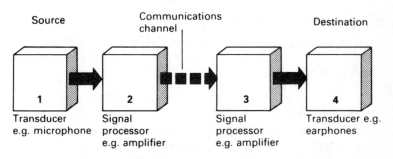

Fig. 16.1 The building blocks of a communications system

munications system. For example, its range could be improved using an amplifier to boost the strength of the signal from the transducer. Similarly, a signal weakened during transmission could be boosted using an amplifier in the receiver. A more sophisticated system might employ an encoder in the transmitter (and a decoder in the receiver), to give signals suitable characteristics for transmission. Note that the system shown in Fig. 16.1 applies equally well to telecommunications systems which are not electronic. For example, when a honey bee has information about the source of a good supply of pollen, it does a special dance (using its body as a transducer) to encode this information into chemical and visual patterns. Other hive-dwellers decode these patterns into the original information about the distance and direction of the source of the pollen. Fig. 16.1 oversimplifies the process of gathering, transmitting and receiving information (and that is specially true of the dance of the honey bee), but it will serve as a model on which to base our understanding of radio, television and optical communications systems.

16.2 The electromagnetic spectrum

The radio, television and light waves that are used for sending messages from one place to another are part of the electromagnetic spectrum. As Fig. 16.2 shows, this spectrum extends from radio waves to gamma rays. It includes X-rays, infrared and ultraviolet rays, and the visible spectrum comprising red, yellow, green, blue and violet light. Now there are two properties that all these waves have:

(*a*) They all travel at a speed, c, of 300 million metres per second (3×10^8 m s^{-1}) in a vacuum (and very slightly slower in air).
(*b*) They all consist of oscillating magnetic and electric fields – hence the name electromagnetic waves.

The strength of the fields associated with electromagnetic waves varies as shown in Fig. 16.3. This variation is known as a sinusoidal waveform and it has four main characteristics:

(*a*) its amplitude, A, which is related to the strength of the electric or magnetic field;
(*b*) its wavelength, λ, which is the distance between two consecutive parts of the waveform which have the same amplitude, i.e. from peak to peak;

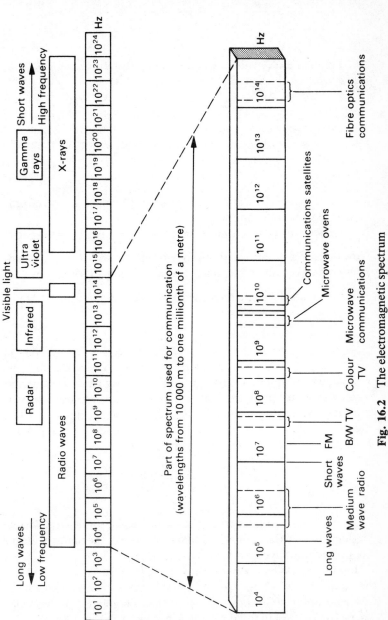

Fig. 16.2 The electromagnetic spectrum

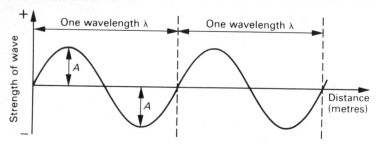

Fig. 16.3   The features of a sinusoidal waveform

(*c*) its frequency, *f*, which is the number of complete waveforms which pass a point in one second; and

(*d*) its speed, *c*, which is the distance the wave moves in one second.

Now there is a simple relationship between the speed of an electromagnetic wave and its frequency and wavelength. This is:

   speed of an electromagnetic wave = frequency × wavelength

or, in symbols,

$$c = f \times \lambda$$

Note that this equation is true for any wave, e.g. sound waves and water waves, not just electromagnetic waves. It can be used to work out the wavelengths of some of the waves shown in Fig. 16.2. Thus medium wave radio waves that have a frequency of, say, 1 MHz (one million hertz) have a wavelength given by

$$\lambda = \frac{c}{f} = \frac{3 \times 10^8}{10^6} = 300 \text{ m}$$

You may have noticed that this wavelength, or its equivalent frequency of 1 MHz, is marked near the centre of the tuning scale on some medium waveband radios.

   It is not necessary to discuss the complex nature of the electric and magnetic fields which are the very essence of electromagnetic waves. But Fig. 16.4 shows that these fields oscillate at right angles to each other and to the direction of the wave. It is usually only the variation of the electric field strength which is shown on waveforms of electromagnetic waves.

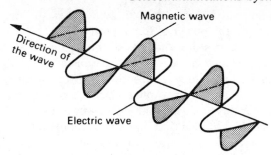

Fig. 16.4 An electric and a magnetic wave make up an electromagnetic wave

16.3 The uses of radio waves

In the electromagnetic spectrum, radio waves extend from a frequency of about 30 kHz (3×10^4 Hz) to more than 3 GHz (3×10^9 Hz). This region is divided into frequency bands as follows.

Band	Frequency	Use
low frequency (LF) or long waves	30 kHz–300 kHz	long distance communications
medium frequency (MF) or medium waves	300 kHz–3 MHz	local sound broadcasts
high frequency (HF) or short waves	3 MHz–30 MHz	distant sound broadcasts, satellite tracking, amateur radio
very high frequency (VHF)	30 MHz–300 MHz	FM sound broadcasts, TV voice communications with spacecraft and space stations (e.g. Soviet Mir space station and US Space Shuttle)
ultra high frequency (UHF)	300 MHz–1 GHz	radar, data to and from interplanetary spacecraft, weather satellites
L-band	1 GHz–2 GHz	ship-to-shore, weather and communications satellites
S-band	2–4 GHz	US Shuttle voice communications, interplanetary spacecraft (e.g. Voyager)
C-band	4–8 GHz	communications satellites
X-band	8–12 GHz	communications and TV satellites, direct broadcast TV
Ku-band	12–18 GHz	communications satellites, direct broadcast TV

Depending on the frequency of radio waves, the shape of the aerial used and the power of the transmitter, radio waves reach a receiver by one or more routes as shown in Fig. 16.5. Surface or ground waves follow the curvature of the Earth. The range of these waves is limited since they are absorbed by poor conductors such as sand, but travel further over water which is a good conductor. The range may be about 1500 km for long waves which have a frequency less than about 300 kHz, but only a few kilometres for very high frequency waves. The sky wave travels upwards from an aerial, but if its frequency is less than a critical frequency of about 30 MHz, it is returned to Earth by the ionosphere. On reaching the ground, the wave is reflected back to the ionosphere, and so on until it is completely attenuated (weakened). The ionosphere consists of several layers of positively charged ions and electrons. These are produced by ionisation of gases in the upper atmosphere by the Sun's ultraviolet radiation. The various layers of the ionosphere extend between about 50 km and 500 km above the Earth, and their intensity, i.e. amount of ionisation, and altitude vary with the time of day, the seasons and the 11-year sunspot cycle. Low, medium and high frequency radio waves have a range of several thousand kilometres and can travel round the world by repeatedly bouncing off the ionosphere.

Radio waves which have frequencies greater than the critical frequency of 30 MHz can penetrate the ionosphere. Indeed, communications with most satellites and interplanetary spacecraft take

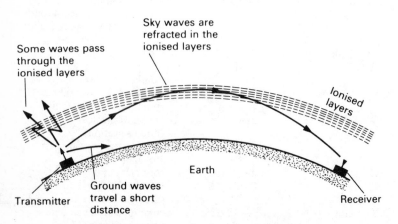

Fig. 16.5 The different paths taken by radio waves

place at frequencies in excess of 1000 MHz (1 GHz). For example, the Intelsat-5 communications satellites re-broadcast TV and telephone messages sent up to them at 11 GHz uplink (i.e. to them) and 14 GHz downlink (i.e. from them). At frequencies of about 100 GHz radio waves begin to be absorbed again by oxygen and other gases in the atmosphere. It is in the 'window' between 30 MHz and 100 GHz that radio astronomers are able to study the radio emissions from distant galaxies and dust clouds (Fig. 16.6).

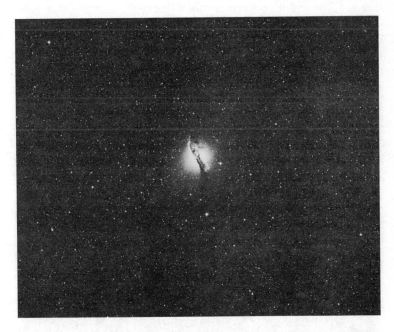

Fig. 16.6 Centaurus-A. This spectacular galaxy is about 15 million light-years away from us and a powerful source of radio emission. The photo shows its ellipsoidal body is crossed by a curious, complicated lane of cool, dense dust which may have been left over after two galaxies collided. Centaurus-A also produces a huge amount of energy at X-ray and optical wavelengths.
Courtesy: Royal Observatory, Edinburgh

Radio waves are produced quite naturally by lightning flashes in the atmosphere of our own planet, and with much greater intensity in the atmospheres of planets Jupiter and Saturn. Also the Sun, the stars and types of galaxies called quasars produce enormous amounts of radio energy. In fact, there's a whole branch of astronomy concerned with trying to understand the processes in deep space which give rise to these radio waves. Radio astronomers use large dish aerials like the one shown in Fig. 16.7 to receive radio waves from space in the search for a better understanding of the kind of universe we live in. Some astronomers use monitoring equipment of this kind to 'listen in' to radio energy from deep space in the hope of receiving messages from distant civilisations! But this background of radio waves from space is not just of academic interest, for it does sometimes interfere with regular radio communications on Earth. Indeed, the Sun is a particularly strong source of 'radio interference', especially when sunspots become visible on its surface during the 11-year sunspot cycle.

Fig. 16.7   The Lovell Radio Telescope in Cheshire, England, still operational after 30 years and now a listed building. This telescope is linked to six other radio telescopes by a computer-controlled landline to provide high resolution maps of cosmic radio sources. The telescope is dedicated to Sir Bernard Lovell, the pioneer of radio astronomy in the UK.
Courtesy: Science Centre and Tree Park, Jodrell Bank

16.4 AM and FM modulation

It is rather difficult to produce radio waves at audio frequencies, i.e. at the frequencies of speech and music. Radio waves at these frequencies require a lot of electrical energy, and larger transmitter aerials are needed as the wavelength of the radio waves increases. Consider the dipole (or Yagi) aerial shown in Fig. 16.8 which consists of two horizontal or vertical conducting rods or wires fed with radio energy from the centre. A vertical dipole like this radiates energy equally in all directions. Now if this aerial is to operate at optimum efficiency, each rod must have a length one quarter of the required wavelength ($\lambda/4$). Thus at a frequency of 10 kHz, each rod of this dipole aerial must be about 7.5 km long since

$$\lambda/4 = c/4f = 3 \times 10^8 \text{ m s}^{-1}/4 \times 10^4 \text{ s}^{-1} = 7500 \text{ m}$$

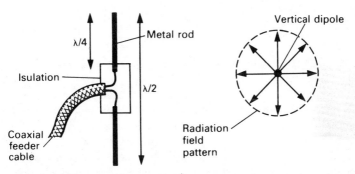

Fig. 16.8 The dimensions and radiation pattern of a dipole aerial

But there is one advantage of low frequency radio waves: they travel quite well through water so they are used for communicating with submarines via long aerials stretched across the surface of islands.

For frequencies above 10 kHz, it becomes increasingly easier to produce radio waves. And with the increase in frequency, shorter aerials can be used for both transmission and reception. Thus data transmitted from a weather satellite at a frequency of 135 MHz (wavelength 2.22 m) can be received on a dipole aerial made of two rods each 55 cm long. A process called **modulation** is used to enable these high frequency radio waves to carry audio frequency information. There are two main processes for modulating a radio frequency carrier wave: amplitude modulation (AM) and frequency modulation (FM).

In **amplitude modulation**, the amplitude of the carrier wave is made to follow the variations in the amplitude of the audio frequency wave as shown in Fig. 16.9. The **modulation depth**, m, of this AM radio wave is defined as a percentage as the ratio (A/B) $\times$ 100%. If this ratio exceeds 100%, the audio frequency message will

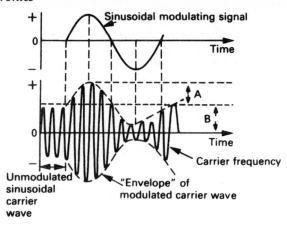

Fig. 16.9 An amplitude modulated carrier wave

be distorted. Too low a modulation depth produces poor quality sound at the receiver. A value of 80% is satisfactory.

In **frequency modulation**, the frequency of the carrier wave is made to follow the amplitude of the audio frequency waves as shown in Fig. 16.10. When the amplitude of the audio frequency wave is zero, the carrier wave has a particular frequency (the frequency marked on the tuning scale of an FM receiver). An increase in the amplitude of the audio frequency wave in the positive direction produces a slight increase in the frequency of the carrier wave. An increase in amplitude in the negative direction produces a decrease in the frequency of the carrier wave. Note that the amplitude of the FM wave remains constant, which allows the transmitter to operate at high efficiency.

The main advantage of FM over AM is that the noise level at the receiver is reduced. Any electrical noise, e.g. from lightning and electrical machinery, tends to amplitude modulate the carrier and this appears at the output of the receiver. But an FM receiver is sensitive to the frequency variations of the carrier wave and not to amplitude variations. Therefore most of the electrical noise is eliminated.

16.5 Bandwidth

In general terms, bandwidth is the frequency range of input signals to which an electronic system responds. Thus, you may recall

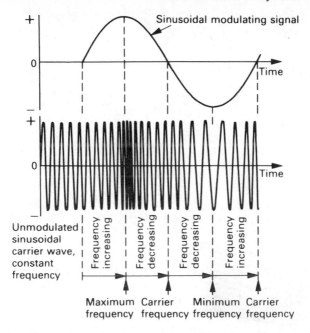

Fig. 16.10 A frequency modulated carrier wave

(Section 8.3) that the bandwidth of an audio amplifier is the range of frequencies it amplifies. Similarly, the bandwidth of a telecommunications system is the capability of that system to transmit signals of different frequencies. As you will see, the bandwidth of a telecommunications system is directly related to the information capacity of that system.

It is possible to show mathematically that when a sinusoidal carrier wave is amplitude modulated (Fig. 16.9), the modulated carrier wave contains three frequencies. Fig. 16.11a shows that one of these frequencies is the original carrier frequency, f_c. The second is the sum of the carrier and modulating frequencies, i.e. $f_c + f_m$. The third is the difference between the carrier and modulating frequencies, i.e. $f_c - f_m$. These two new frequencies are called side frequencies. The sum of the carrier and modulating frequencies is called the upper side frequency, and the difference between the two frequencies the lower side frequency. The bandwidth of this amplitude modulated carrier wave is the difference between the two side frequencies:

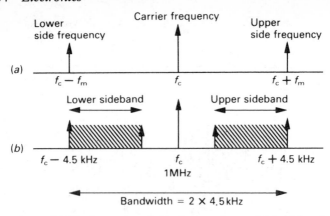

Fig. 16.11 (a) The formation of side frequencies, and (b) sidebands in an AM radio wave

$$\text{bandwidth} = (f_c + f_m) - (f_c - f_m) = 2f_m$$

i.e. twice the modulating frequency.

When the modulating signal consists of a band of frequencies, as in speech and music, each individual frequency produces upper and lower side frequencies about the unmodulated carrier wave. This results in upper and lower sideband frequencies as shown in Fig. 16.11b. Thus, if the maximum frequency of the modulating signal is 4.5 kHz, the bandwidth required is 9 kHz centred on a carrier frequency of, say, 1 MHz – the middle of the medium waveband. The higher the modulating signal bandwidth, the wider the modulated signal bandwidth. It is obvious that the transmission system used must be capable of handling the bandwidth required of it. In order to avoid overcrowding on the AM medium waveband which extends from about 500 kHz to 1.5 MHz, each station is allocated a 9 kHz bandwidth.

Sidebands are also produced when a carrier wave is frequency modulated. But the arrangement of the sidebands is much more complicated than with amplitude modulation. A number of sidebands are produced, not just two, and the way they are distributed and the number of them is complicated. The number of sidebands is determined by the ratio

$$\frac{\text{maximum variation of carrier frequency}}{\text{modulating frequency}}$$

and is known as the **modulation index**.

The maximum variation of carrier frequency is determined by the performance required. For the BBC's FM sound broadcasts on the VHF radio band, this is 75 kHz. If the modulating frequency is 15 kHz (about the maximum transmitted), the modulation index is 75 kHz/15 kHz, i.e. 5, and the number of sidebands is 16 (8 on each side of the carrier). These are all spaced 15 kHz apart so the total bandwidth is 240 kHz. Moreover, the bandwidth is dependent on the magnitude of the modulating voltage whereas in AM the bandwidth is fixed. Thus the bandwidth required by FM is much greater than that required by AM, i.e. 240 kHz for FM and 30 kHz for AM if the modulating signal has a frequency of 15 kHz. This higher bandwidth necessitates the use of higher carrier frequencies which is why FM sound broadcasting is on the VHF band – see Fig. 16.2. Television (Section 16.7) requires an even larger bandwidth: the 625-line system needs a bandwidth of about 8 MHz per channel which is why 625-line TV is accommodated between 610 MHz and 940 MHz of the electromagnetic spectrum. Perhaps you can see the advantage of using light to carry information? Since the frequency of light is about 1000 times greater than the highest frequency radio wave band, communicating with light offers an enormous information-carrying capacity.

16.6 Radio transmitters and receivers

Fig. 16.12 shows the main building blocks needed to make an AM radio transmitter. The carrier wave is generated by the radio frequency (RF) oscillator, A, and amplified by the RF amplifier, B, before being passed to the modulated RF amplifier, C. The audio frequency (AF) signal produced by the microphone is amplified by the AF amplifier, E, and modulator, F, and passed to the modulated RF amplifier, C. An AM radio frequency signal is fed to the aerial via the optional RF amplifier, D.

The AM radio frequency signals can be received using the system shown in Fig. 16.13. An RF amplifier, A, selects a particular AM carrier wave and amplifies it. The selection is done by a tuned circuit (see below) and amplified by an RF amplifier which has a bandwidth of 9 kHz to accept the sidebands. A detector (or demodulator – see below), B, discards the RF carrier wave to leave the required AF signal. An amplified AF signal is then fed to the loudspeaker.

The tuned circuit of an AM radio receiver comprises a coil of wire, usually wound on a ferrite rod, and a capacitor connected in parallel with the coil as shown in Fig. 16.14a. This building block

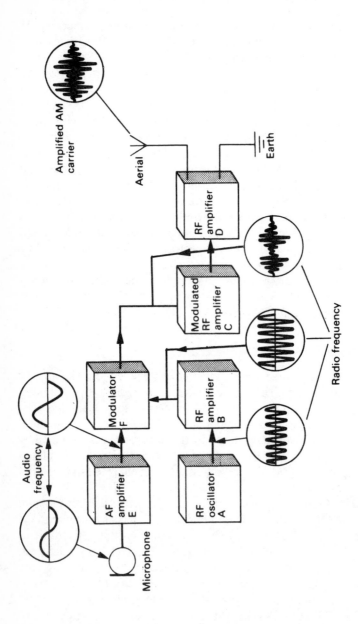

Fig. 16.12 The building blocks of an AM transmitter

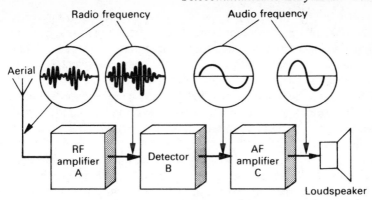

Fig. 16.13 The building blocks of a tuned radio frequency receiver

resonates at a characteristic frequency determined by the values of the inductance, L, of the coil and the capacitance, C, of the capacitor. Inductance is measured in units of henries (H) while capacitance (Chapter 6) is measured in farads (F). In an AM radio receiver, the inductance of the coil would be a few microhenries and the capacitance of the capacitor a few hundred picofarads. By using a variable capacitor, different stations can be tuned in by making the tuned circuit resonate at different frequencies. The resonant frequency, f_0, is given by the equation

$$f_0 = 1/(6.28\sqrt{LC})$$

which indicates that increasing the values of C and/or L reduces the frequency at which the circuit resonates, i.e. increases the wavelength of the AM waves selected. At the resonant frequency, the p.d. developed across the tuned circuit has a maximum value as

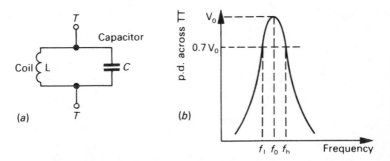

Fig. 16.14 (a) A parallel tuned circuit, and (b) the p.d. across it at its resonant frequency

shown in Fig. 16.4b; negligible p.d.s are developed in the tuned circuit for all the other frequencies that the aerial is picking up. Thus it is the resonant frequency that is amplified by the circuit that follows it. Actually, the tuned circuit does not select just one frequency but a narrow band of frequencies as shown in the graph. Obviously if the tuned circuit is to 'separate' closely-spaced stations, the bandwidth must be small – but not too small since the carrier wave from a particular station requires a certain minimum bandwidth in order to carry a full range of audio frequencies – see Section 16.5. In Fig. 16.14b, the bandwidth is defined as the difference between the higher and lower frequencies, f_h and f_l, where the p.d. across the tuned circuit has fallen to 0.7 of the maximum p.d.

The process of detection or **demodulation** of an AM radio wave can be achieved with the circuit shown in Fig. 16.15. During each positive part of the radio frequency signal, diode D_1 passes a current which charges up capacitor C_1. As the positive peaks subside, the capacitor discharges through R_1. If the capacitor voltage remains close to the peak voltage until the next peak of the radio frequency signal arrives, only the audio frequency envelope of the AM signal is passed by the detector; i.e. the detector is a low-pass filter. This means that the time constant $R_1 \times C_1$ must be between one period of the RF signal (say, 10^{-6} s for a radio frequency of 1 MHz) and one period of the AF signal (say, 10^{-3} s). If the time constant is too large, C_1 discharges too slowly and the output voltage does not follow the AF signal, i.e. the detector is insensitive. And if the time constant is too small, the RF signal passes through the demodulator. Typical values are $C_1 = 0.01$ μF and $R_1 = 10$ kΩ giving a time constant of 10^{-4} s.

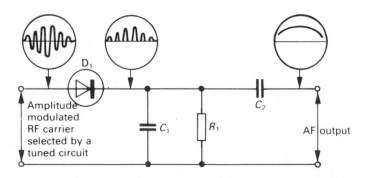

Fig. 16.15   The principle of demodulation

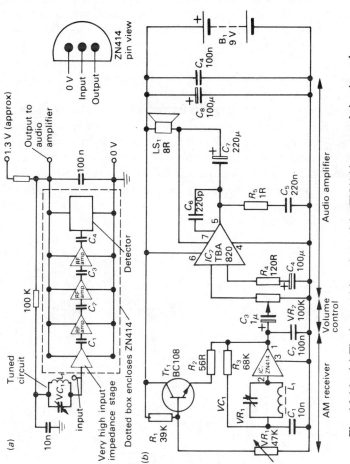

Fig. 16.16 (a) The internal functions of the ZN414 integrated circuit, and (b) the design of an AM radio receiver based on it

An AM radio receiver can be built using a specialised integrated circuit such as the Ferranti ZN414 device shown in the circuit of Fig. 16.16. The ZN414 is a 3-pin transistor-like IC which is designed to amplify the p.d. generated across the tuned circuit by the radio frequencies to which the circuit resonates. After amplification, the radio signals are detected inside the ZN414, after which the audio frequency signals are amplified by a second integrated circuit, IC_2, so that a low-power loudspeaker can be operated. The TBA820 is a low-power audio frequency amplifier. The ZN414 requires a supply e.m.f. of about 1.3 V. In this circuit, this e.m.f. is provided by a simple circuit based on transistor Tr_1 and resistors VR_1, R_1, R_2 and R_3.

A simplified systems diagram for an FM transmitter is shown in Fig. 16.17. The audio output from the microphone is amplified by the radio frequency amplifier, A, and fed to the RF oscillator, B. This produces a frequency modulated signal which is fed to the RF amplifier, C, and then to the aerial. Note that the RF amplifier increases the amplitude of the signal but does not alter the frequency changes.

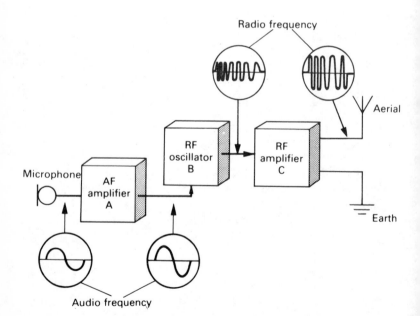

Fig. 16.17 The building blocks of an FM transmitter

These FM radio signals can be received with the system shown in Fig. 16.18. This is basically the same as for the reception of AM radio signals except that the demodulator has to respond to the variations of the carrier frequency instead of variations of carrier amplitude.

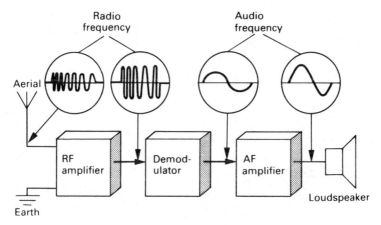

Fig. 16.18 The building blocks of an FM receiver

16.7 Television basics

Television is the transmission and reception of pictures and sound by very high frequency (VHF) radio waves (see Fig. 16.2). A simplified diagram of a television broadcast system is shown in Fig. 16.19. In the television studio, or during outdoor broadcasts, one or more television cameras produce electrical signals of a moving visible scene. A video amplifier amplifies these signals and sends them along a line to the transmitting station. At the same time, sounds associated with the scene are converted into electrical signals by a microphone and these, too, are amplified along with the video signal. At the television transmitting station, two carrier waves are modulated, one with the video information and one with the sound information. The transmitting aerial radiates a single carrier wave modulated with both video and sound information. A receiving aerial picks up this wave and amplifies it before demodulating the vision and sound components. The two signals are separated out. The demodulated sound signal passes to an AF amplifier before being passed to a loudspeaker which recreates the original sound as accurately as possible. And the demodulated

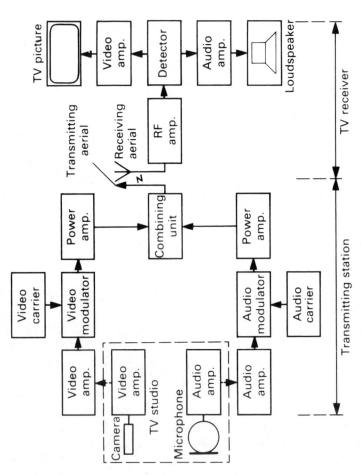

Fig. 16.19 A very simplified television broadcast system

video signal is amplified by a video amplifier before being passed to a cathode ray tube in the TV receiver to recreate an image of the original scene.

A black-and-white television camera combines optics and electronics to produce electronic signals representing the variations in brightness of the scene. The basis of one type of camera is the **vidicon** tube shown in Fig. 16.20. It produces a narrow beam of electrons aimed at a photosensitive material on which a lens system focuses an optical image. In front of the target is a transparent aluminium film. The resistance between this film and any point on the back of the target immediately behind it depends on the brightness of the image at that point. The brighter the image, the lower the resistance at that point. The variation in brightness of the image is 'read' using an electron beam which is made to move back and forth over the target. During this scanning process, the electron current varies in the circuit comprising the electron beam, the target, resistor R, the power supply and the electron gun. The strength of the electron current thus follows the brightness of the image, and resistor R turns these current variations into voltage variations which comprise the video signal. A different type of camera, one based on semiconductors rather than thermionic emission, is the CCD camera. The CCD (charge-coupled device) comprises an array of metal-oxide semiconductors on a silicon chip. It is much more sensitive that the vidicon tube and it is also able to store images.

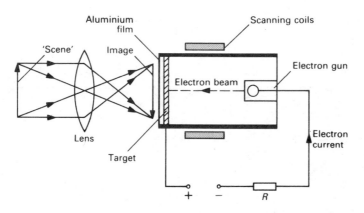

Fig. 16.20 The principle of a vidicon tube

264   *Electronics*

In the vidicon tube, the electron beam is made to scan the target by suitable signals passing through the scanning coils wound round the tube. The spot of electrons is first positioned at the top left of the target and rapidly moved horizontally as shown in Fig. 16.21a. At the end of the line, the spot is moved rapidly back to the left side of the target but positioned just below the starting point of the first line. This very rapid return of the spot is called fly-back. A second line is now traced and the process repeated until the spot has reached the bottom right-hand corner of the target. The spot is then returned to the top left-hand corner of the screen and a second picture traced out. Thus scanning produces a picture made up of a set of parallel lines called a **raster**. The system that deflects the spot horizontally is called the field scan, and that which moves the spot downwards at a much slower rate is called line scan. The deflection is produced by magnetic fields generated by currents which have a sawtooth shape as shown in Fig. 16.21b. Scanning is a common way of building up pictures and is used in the cameras on weather satellites, and on interplanetary spacecraft such as *Voyager 2* which sends us detailed pictures of planets during its journey through the solar system.

In television, the raster must have at least 500 lines if a 'grainy' picture is to be avoided, and the complete scans should follow each other fast enough to allow the eye's persistence of vision to see a moving picture without any significant flicker. Individual pictures presented to the eye faster than 16 times per second are needed. In

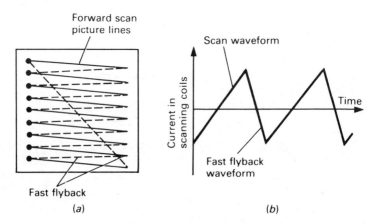

Fig. 16.21   (a) How scanning produces a picture; (b) driving the scanning coils with a sawtooth waveform

practice, in order to avoid 'hum bars' in the picture, the number of pictures per second has to equal the frequency of the a.c. mains. Thus in the UK, 50 pictures per second, and in the USA 60 pictures per second are transmitted. The number of lines used in television systems varies from country to country and has changed over the years. The original BBC and ITA channels on the 30–300 MHz band used 405 lines; now the transmissions are in the 300 to 3000 MHz band and use 625 lines. Whereas the original 405-line system used amplitude modulation for both sound and vision signals, the 625-line system uses amplitude modulation for vision and frequency modulation for sound.

In a TV receiver the picture is displayed on the screen of a cathode ray tube (CRT). Fig. 16.22 shows that the TV tube is similar in design to the CRT used in an oscilloscope (Section 4.4), except for the use of deflection coils instead of X and Y deflection electrodes. The tube is an airless glass envelope, narrow and cylindrical at one end and flared out at the other end to form a viewing screen. Electrons are produced by heating a filament at the narrow end, and these are accelerated and focused into a narrow beam by an electron gun before being 'fired' along the axis of the tube towards the positive anodes. The horizontal and vertical deflection coils clamped round the neck of the tube move the beam back and forth across the screen to build up a picture as a set of lines on the screen.

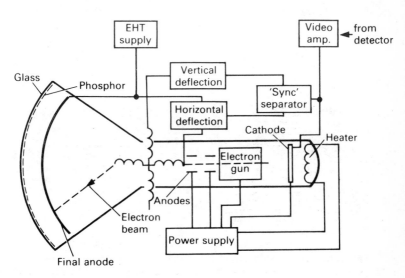

Fig. 16.22 The principles of a cathode ray tube in a TV receiver

Where the electrons strike the phosphor, light is produced, the brightness of which is determined by the energy of the electrons. External controls on the receiver enable the intensity of the electron beam and therefore the overall brightness of the screen to be adjusted.

Because a TV picture contains a lot of information, TV needs a bandwidth of about 11 MHz. In order to save 'radio space', this rather large bandwidth is halved to 5.5 MHz by a technique called interlaced scanning. Each complete picture is divided into two frames (or fields) which are scanned and transmitted one after the other and then reassembled at the TV receiver. Thus after the beam has scanned half the picture comprising 312.5 lines in 1/50 s, it returns to scan the intervening lines. Therefore, although 50 frames (in the UK) are transmitted per second, only 25 complete pictures are transmitted per second. The vision signal therefore contains only half the picture information thereby reducing the bandwidth by half to 5.5 MHz. However, to transmit this bandwidth using amplitude modulation requires an overall bandwidth of 11 MHz to accommodate the sidebands. In practice, it is found that a satisfactory picture is received if the whole of the upper sideband is transmitted and a part (a vestige) of the other. In Fig. 16.23, the frequencies are shown relative to the allocated vision carrier frequency. The audio carrier wave is given a 250 kHz bandwidth above the video carrier wave and above the maximum frequency of the upper sideband. And the lower sideband is restricted to 1.25 MHz

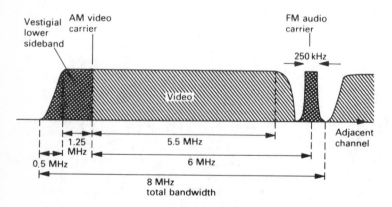

Fig. 16.23 The bandwidth requirements of a 625-line TV channel

below the video carrier frequency. This gives an 8 MHz channel bandwidth for each TV station.

Finally, to end this simple introduction to television, note that the electron beam that scans the picture in the vidicon tube must be in exactly the same position at all times as the electron beam that scans the screen in the picture tube. This is achieved by transmitting synchronising pulses along with the picture information. Both line and frame synchronising pulses are added to the video signal during the flyback times when the line is blanked out. In the TV receiver, these pulses are separated out from the video signal and used to trigger line and frame time-base circuits which supply the currents to the deflection coils that position the spot of light on the TV screen.

16.8 Colour television

When colour television was introduced in the UK in 1967, consideration had to be given to owners of monochrome receivers so that they could continue to receive a normal monochrome picture. It was also necessary for a colour receiver to display a good monochrome picture, for example of early films. A TV signal which allows both a monochrome and a colour set to receive pictures is known as a compatible signal. It is produced in a TV camera in two distinct parts: a luminance signal and a chrominance signal.

The luminance part of the signal contains the brightness information just as it does for a monochrome-only system. The chrominance part of the signal contains the extra information destined for a colour picture. Thus a monochrome receiver uses only the luminance part of the signal while a colour receiver makes use of both parts of the signal. World-wide, there are three methods of providing a compatible signal from a TV camera.

(a) The National Television Systems Committee (NTSC) system which was developed and used in the US in the early 1950s, and later also adopted in Canada, Japan and Mexico.
(b) The Phase Alternate Line (PAL) system developed in West Germany from the NTSC system, and later also adopted in the UK and other European countries.
(c) The Sequential Colour and Memory (SECAM) system developed in France, and later also adopted in East Germany, USSR and other countries in Europe and North Africa.

In all three systems, the luminance and chrominance information, including the line and frame synchronising pulses, form a compatible video signal which then amplitude modulates a video carrier signal for transmission. This transmitted signal has to have the same bandwidth, i.e. 8 MHz, as a 625-line monochrome signal. But since the eye is more sensitive to changes of light intensity than to changes of colour, the bandwidth required for chrominance information is less than for luminance information. In the PAL system, the signals representing chrominance and luminance information are interleaved at the transmitter by an encoder. The luminance information occupies the same bandwidth as for monochrome transmissions, and the chrominance information amplitude modulates a subcarrier 4.43 MHz from the luminance carrier, its sidebands extending about 1 MHz on either side as shown in Fig. 16.24. At the receiver, a decoder separates the chrominance and luminance information.

In a colour camera, three vidicon tubes have to be used, one for each of the three colours, red, green and blue as shown in Fig. 16.25a. These are primary colours and can be combined in suitable proportions to form white light. Various combinations of pairs of these primary colours produce complementary colours. Thus when red and green light are mixed, yellow is seen by the human eye; green and blue give cyan; red and blue, magenta. In the camera, the primary colours are extracted from the scene using suitable light filters in front of each of the camera tubes. The camera tubes are scanned in synchronism using internal electron beams which produce three separate voltages, V_R, V_G and V_B, representing the

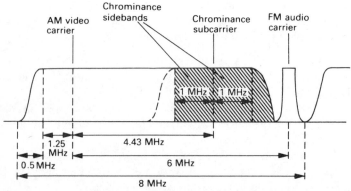

Fig. 16.24   How the chrominance signal is accommodated in a transmitted video signal

colour information. Next, encoding circuits provide a chrominance signal for amplitude modulating the video carrier, and a luminance matrix produces a voltage V_L, which represents the brightness variations in the scene, and is exactly the same signal as the video signal generated in the monochrome camera shown in Fig. 16.19. The luminance matrix generates the luminance signal by adding together suitable proportions of the V_R, V_G and V_B outputs from the colour camera. The chrominance signal produced by the encoder is obtained as three colour-difference signals; these are $V_R - V_L$ (red minus luminance), $V_G - V_L$ (green minus luminance), and $V_B - V_L$ (blue minus luminance). Only two of these colour-difference signals, the red minus and blue minus, are used to modulate the sub-carrier frequency, the green minus signal being extracted at the receiver.

Fig. 16.25b shows that the colour TV receiver has a similar design to the monochrome receiver except for three main additions: a block to decode the chrominance signal, a tricolour tube containing 'red', 'green' and 'blue' electron guns, and a convergence unit. The signals from the chrominance decoder turn the three guns on and off by varying amounts to change the colour on any part of the screen. For example, if yellow is to be produced, the 'red' and 'green' guns would be turned on and the 'blue' gun switched off. To produce

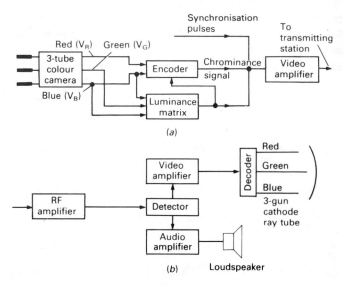

Fig. 16.25 (a) The functions of a colour TV camera; (b) the functions of a colour TV receiver

white, all three guns have to be switched on. The convergence unit makes sure that the three coloured rasters of the colour tube are in register or 'converged' by feeding suitably shaped current wave-forms to convergence coils round the neck of the tube. The sound signal is dealt with in exactly the same way as in a monochrome receiver.

There are different designs for the colour tube. One design, shown in Fig. 16.26, has a screen coated with many thousands of tiny red, green and blue phosphor dots arranged in triangular groups (the delta gun tube). Between the guns and the screen is a **shadow mask** consisting of a metal sheet pierced with about half a million tiny holes. Since all three electron guns scan the screen under the control of the same deflection coils, the shadow mask ensures that each beam only strikes dots of one phosphor. Thus electrons from the 'blue' gun can only strike the blue phosphor dots. When a particular triangle of dots is scanned, it may be that only the 'green' and 'blue' electron beams are intense and the 'red' is weak. In this case the triangle would emit blue and green light and the group of phosphor dots appears cyan. There are other designs of tube, notably the Trinitron tube invented by Sony (Japan). In this tube, three beams are emitted from one electron gun, and an 'aperture grill' produces vertical phosphor stripes on the screen.

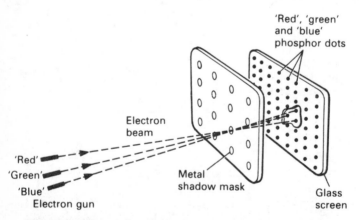

Fig. 16.26 How a shadow mask works in a colour TV receiver

16.9 Fibre optics communications

For a long time we've used light to send messages. Our forefathers lit beacons when invaders threatened. A hand-held mirror, the heliograph, was first used by the ancient Greeks to reflect the Sun's rays and flash coded signals over great distances. And lighthouses and traffic lights use light to warn us of danger. But fibre optics communications is an altogether more sophisticated way of sending messages from one place to another: it makes use of long thin glass fibres along which information is sent as pulses of laser light. Fibre optics communications is already in use and it holds great promise for the future of telecommunications systems. But why should optical fibres be so superior to conventional copper cables?

Well, for one thing cables made from optical fibres are cheaper, lighter and easier to install than copper cables. Furthermore, they are completely free from electromagnetic interference since data on a light beam cannot be corrupted by electrical machinery, thunderstorms and 'noisy' power lines. Thus there is no interference or 'cross-talk' between neighbouring fibres, a quality which also means that signals carried by optical fibres are much less liable to be detected compared with electrical signals in copper cables, i.e. the information is effectively secure from eavesdroppers. Safety, too, is an important reason for using optical fibres since broken fibres are not a fire hazard as the escaping light is harmless. But perhaps the strongest justification for using optical fibres is their potential for carrying considerably more information than copper cables. A glance at Fig. 16.2 shows why. Since light waves have frequencies about 10 000 times higher than the highest frequency radio waves, considerably greater bandwidth (Section 16.5) is available. Indeed, conventional copper cables are hard pressed to keep up with the mounting speed of development in communications and information technology. However, before fibre optics communications could become a reality, we had to wait for two high technology inventions: the laser and low-loss fibre optic cable.

By 'low loss' is meant glass so pure you could see through a 35 km thick block of it as clearly as through a window pane! Such high purity means that information travels through optical fibres for long distances without having to be repeatedly amplified on the way – it is said to have low attenuation. The raw material for such optically clean fibres is a special kind of sand called silica. An optical fibre is a solid rod of silica, finer than a human hair and surprisingly flexible. It is manufactured in the cleanest of atmospheres to ensure that no speck of dust or fingerprint can mar its purity.

The basic construction of an optical fibre is shown in Fig. 16.27. It comprises a glass core totally enclosed by a glass cladding. A plastic coating covers the cladding and core to prevent dust and moisture from reaching the glass core. Light (actually, it's infrared – see

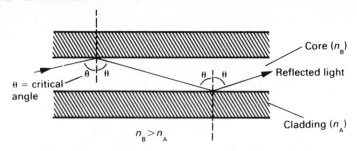

Fig. 16.27   How light travels through an optical fibre

below) doesn't just travel along the core in any old way; it is guided along the core by being reflected back from the outer cladding to the core, bouncing along from side to side of the core. No light is lost as it bounces from the cladding; at each bounce it is all reflected back to the core. What makes it do this is the relative optical properties of the core and the cladding. The cable is designed so that the refractive index of the core is higher than the refractive index of the cladding. This ensures that light which meets the boundary between the core and cladding at an angle greater than a certain 'critical' angle is totally reflected back into the core. This is called **total internal reflection**.

Fig. 16.28 shows a fibre optics cable being laid between the Isle of Wight and mainland Britain. This is an armoured cable designed both to protect the bundle of fibres from contact with moisture and chemicals, and to strengthen it.

Readers who know some basic optics may recall that a simple equation is used to calculate the critical, i.e. minimum, angle, θ, at which the light must meet the boundary if it is to be totally internally reflected. It is

$$\sin \theta = n_A/n_B$$

where n_B is the refractive index of the glass of the core and n_A is the refractive index of the cladding. Thus if $n_A = 1.43$ and $n_B = 1.45$, $\sin \theta = 1.43/1.45 = 0.9862$. Therefore, $\theta = 80°28'$.

For angles of incidence less than θ the light passes into the cladding and is lost. Of course, glasses with different refractive indices give different values of critical angle.

Three different types of optical fibre are in common use as shown in Fig. 16.29. First, there is the stepped-index mono-mode fibre.

Fig. 16.28 Optical fibres being laid between the Isle of Wight and mainland Britain
Courtesy: British Telecom

This has a very narrow core through which light follows one path, i.e. mode. Since the core's diameter is of the order of the wavelength of the light used, all the light waves from the source follow the same path along the fibre. This results in a high data transmission rate. Indeed, these fibres have the greatest information carrying capacity, i.e. bandwidth, of all types of fibre and are used for the longest transmission distances. They are, however, quite expensive to manufacture and their use is generally limited to very high-capacity large-bandwidth systems such as undersea cables where the expense is justified by the high return of earned income.

For data links that require a relatively low bandwidth and information-carrying capacity, stepped-index multi-mode fibres are generally used. Like the stepped-index mono-mode fibre, this fibre has a sudden change of refractive index at the core–cladding boundary, but it has a wider core diameter than the single-mode fibre. The wider diameter means that light waves can take different paths along the core and so take different times to reach their destination. This produces distortion known as transit-time disper-

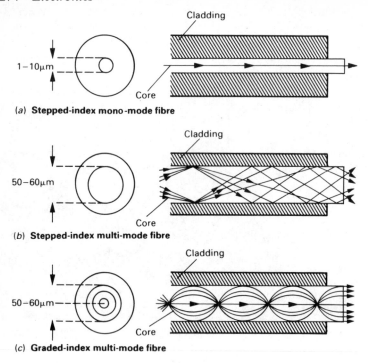

Fig. 16.29   The construction of three types of optical fibre

sion which limits the rate at which data can be modulated, i.e. successive pulses of light merge into each other and cause distortion of the information being carried.

For intermediate bandwidth and capacity systems, graded-index multi-mode fibres are used in which the refractive index decreases gradually from the centre of the core to the cladding. This ensures that individual light waves are gradually bent in the core instead of being reflected at the core–cladding boundary. Although waves with a larger critical angle travel further than those with smaller angles, the decrease of refractive index also allows a higher speed of energy propagation. As a result, all waves reach their destination at virtually the same time, thus greatly reducing transit-time dispersion.

Two light sources (or transmitters) are eminently suitable for the job of sending pulses of light down these slender optical fibres:

light-emitting diodes (LEDs) and injection-laser diodes (ILDs). Both sources generate light when excited by electricity, and they are the only sources of light capable of being switched on and off fast enough to be modulated by low power analogue or digital signals. Their physical dimensions are compatible with optical fibres and they have the reliability and long life needed in telecommunications systems. Gallium arsenide (GaAs – see Section 13.10), gallium aluminium arsenide (GaAlAs) and gallium indium arsenide phosphide (GaInAsP) are the materials used in their construction. They convert electricity into infrared energy efficiently, and the special glass used in fibre optics is more transparent at infrared wavelengths. Infrared LEDs are suitable for use with stepped-index multi-mode fibres since they emit a relatively wide beam of light with a fairly large spectral bandwidth. ILD sources radiate a much narrower beam of infrared with a much narrower spectral width. Thus ILD sources are ideal for use with stepped-index mono-mode fibres. Furthermore, ILDs can launch between 0.5 mW to 5 mW of infrared power into a fibre, compared with the smaller 0.05 mW to 0.5 mW for an LED. And ILDs can be modulated at frequencies of over 500 MHz compared with about 50 MHz for LEDs. Gallium aluminium arsenide ILDs and LEDs generate infrared in the 0.8 μm to 0.9 μm range, while gallium indium arsenide phosphide devices generate infrared in the 1.3 μm to 1.6 μm range where attenuation and dispersion by fibres is very low.

A photodiode is generally used to convert the modulated infrared light back into electrical signals at the end of the fibre. The photodiode is reverse-biased so that when it absorbs infrared, a small current flows between its cathode and anode terminals. The current is virtually proportional to the amount of light it absorbs. Photodiodes are generally based on silicon. Infrared emitting diodes and photodiodes are available as spectrally-matched pairs: the LED emits maximum infrared radiation at the wavelength to which the photodiode is most sensitive. This wavelength is typically 0.9 μm.

A practical set-up in which radio audio signals are transmitted using fibre optics is shown in Fig. 16.30. The radio receiver is the circuit shown in Fig. 16.16 and is based on the ZN414 integrated circuit 'radio on a chip'. An audio amplifier based on the TBA820M IC provides audio signals for modulating the optical output from an infrared light-emitting diode, LED$_1$. This LED is contained in a fibre optics housing and launches the optical signals along a fibre optics cable. At the destination, a photodiode converts the infrared

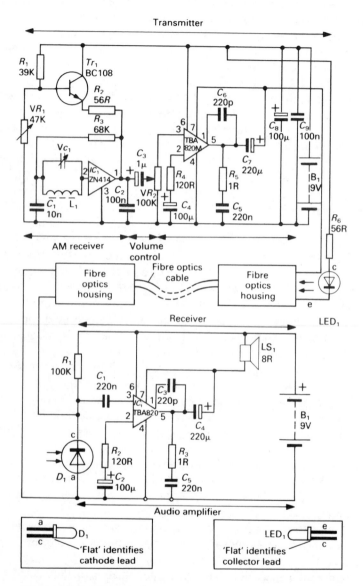

Fig. 16.30 Sending analogue signals through an optical fibre

signals back into electrical signals which are amplified by a second audio amplifier to operate the loudspeaker, LS_1. Note that LED_1 must have the resistor R_6 in series with it, and the photodiode is operated in the reverse-biased mode using a 100 kΩ series resistor. This telecommunications circuit could be used as a voice link by replacing the radio with a microphone, or it could be used to send digital data from one microcomputer to another. In addition to the components shown on the diagram of the radio receiver (see Fig. 16.16) the following components and materials are needed:

* fibre optics housing: 2 off; called a 'sweetspot' housing, this accepts a trimmed fibre optics cable
* fibre optics cable: single core, polymer type; core diameter 1 mm, overall diameter, 2. 25 mm
* LED_1 and D_1: infrared source type OP160, and infrared sensor type OP500 for pushing into sweetspot housing
* second audio amplifier in receiver; components as per circuit diagram

16.10 Pulse code modulation (p.c.m.)

It is not usual to send messages along optical fibres using amplitude modulated infrared as in the above experiment. Commercial optical communications systems use digital techniques to produce pulses of infrared coded in some way to carry the information required. There are several ways of producing digitally coded information. One of these, called pulse code modulation, is used in radio as well as optical communications systems. Indeed, it is the preferred method for communicating with spacecraft and communications satellites.

In pulse code modulation an analogue signal is sampled at intervals as shown in Fig. 16.31 to give a pulse amplitude modulated (p.a.m.) signal. To ensure that the information can be transmitted and recovered without undue reduction of intelligibility, the frequency at which the analogue signal is sampled has to be at least twice the bandwidth of the signal. Now although the ear responds to signals in the frequency range of 20 to 20 000 Hz, voice is intelligible in the frequency range 300 to 3400 Hz. Thus a sampling frequency of 8 kHz is used to accommodate voice transmissions. Generally the higher the sampling frequency, the better the copy of the original analogue signal that is recovered at the destination.

The amplitude of each sample is measured using a scale of levels.

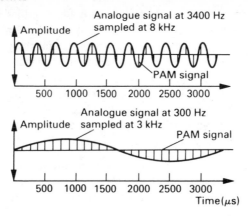

Fig. 16.31 The sampling of an analogue signal to produce a pulse amplitude modulated signal

This measurement is then encoded into a binary equivalent using an analogue-to-digital converter (Chapter 15), and this binary equivalent is known as a pulse code modulated signal. The technique is shown in Fig. 16.32. Here a three-bit binary code is used to represent eight possible (0 to 7) levels of the sample taken at equal intervals of time, t_1 to t_7. The binary equivalent of each sample is then transmitted as a pulse waveform. At the destination a digital-to-analogue converter decodes the digital signal back into an analogue one. The resolution of the sample, i.e. the accuracy with which it represents the magnitude of the analogue signal at the point the sample is made, is determined by the number of bits per sample. Thus if an eight-bit binary number is used, 256 levels can represent each sample.

There are several advantages of using pulse code modulation. First, it is a powerful method of overcoming noisy environments since it is only the presence or absence of a pulse that is needed to construct the original signal, not its shape. In this way it is possible to glean information from extremely weak signals received from the Voyager spacecraft transmitting from the edge of the solar system. Second, in a long distance communications link, a signal is apt to lose energy and be lost. The power of the signal is regenerated en route using repeaters. However, in an analogue signal, not only the signal but also the noise is amplified. In the case of a digital signal, it is regenerated as a pure, clean signal since only the on/off states of

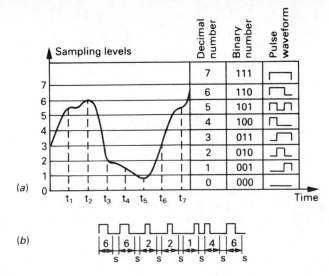

Fig. 16.32 (a) Assigning a 3-bit binary code to each sample; (b) the transmitted pulse waveform

the pulse are identified. Third, once a signal has been digitised it is compatible with other digital signals. Thus digitised television, music and voice signals may be multiplexed (Chapter 11) and transmitted as one signal.

Through British Telecom's use of System X, the British telephone network is fast being digitised. Electronic logic gates (Chapter 9) in computer-controlled exchanges have replaced electromechanical relays for handling digital signals from telephones and computers. Finally, you may remember reading in Section 1.2 that both the compact disk and video disk use digital methods for storing and retrieving sound and pictures.

Glossary

A The symbol for the unit of electrical current, the ampere

A The symbol used in the hexadecimal system for the decimal number 10. Thus the hexadecimal number A4 is equal to the decimal number $10 \times 16 + 4 = 164$

absolute temperature A temperature measured on the absolute, or kelvin (K), scale which has the melting point of ice at -273 °C, and the same interval between degrees as on the Celsius scale. Thus an absolute temperature of 290 K is equal to $290 - 273 = 17$ °C

address A memory location in a computer usually identified by a decimal or hexadecimal number. Thus the hexadecimal address FE61 is equivalent to the decimal address 65121

aerial An arrangement of electrical conductors, usually placed in an elevated position (e.g. on or between masts), which transmits and/or receives radio signals

AF (audio frequency)

alphanumerics The use of seven-segment displays, liquid crystal displays, and other optoelectronics devices, to display numbers, letters and some punctuation marks and mathematical symbols

American standard code for information interchange (ASCII) A set of 128 characters comprising numbers, letters, punctuation marks and symbols, each represented by a 7-bit binary word, which was introduced to facilitate the exchange of information between a computer and other processing equipment

ammeter An instrument for measuring the strength of an electric current in units of amperes, milliamperes or microamperes

ampere A flow of electric charge in a circuit equal to 1 coulomb per second

amplifier A device that increases the voltage, current or power of a signal

amplitude The strength of a signal measured by its maximum value

amplitude modulation (AM) A method of sending a message on a radio or light wave by varying the amplitude of the wave in response to the frequency of the message

analogue signals Signals, e.g. amplitude modulated radio waves, that respond to or produce a continuous range of values rather than specific values

analogue-to-digital converter (ADC) A device that changes an analogue signal into a digital signal. Thus an ADC is used when inputting a temperature reading from a thermometer into a computer

AND gate A building block in digital logic circuits that makes the following logical decision: it only produces an output of logic 1 when all of its inputs are at logic 1

anode The terminal of a device (e.g. a diode) towards which electrons flow

arithmetic and logic unit (ALU) That section of a microprocessor which makes use of logic gates to add and subtract numbers

artificial intelligence (AI) The rapidly developing study of machines which perform tasks that could be considered intelligent if done by humans

ASCII (American standard code for information interchange)

astable A circuit designed to produce a continuous signal that has a rectangular or square waveform. Astables have numerous applications, e.g. as 'clocks' in digital circuits and in alarm and monitoring systems

atom the smallest particle of a chemical element that can exist alone or in combination with other atoms

audio frequency (AF) Any regularly repeating signals which are in the frequency range of human hearing, i.e. from about 20 Hz to 20 kHz

automation Any device or system that takes the place of humans in carrying out repetitive and boring jobs

B The symbol in the hexadecimal system for the decimal number 11. Thus the hexadecimal number 9B equals the decimal number $9 \times 16 + 11 = 155$

bandwidth The range of frequencies contained in a signal, or to which a device responds. Thus a device that responds to the frequency range 0.5 kHz to 4 kHz has a bandwidth of 3.5 kHz

bargraph display An instrument readout which indicates the value of something by the length of a row of glowing light-emitting diodes, or the active elements of a liquid crystal display

base One of the three terminals of a bipolar transistor into or out of which a small base current flows and controls a much larger collector current

Basic (beginners all-purpose symbolic instruction code) An introductory high-level computer programming language

BCD (binary-coded decimal)

bias The current or voltage which is applied to part of a circuit to make the circuit function properly

binary-coded decimal (BCD) Any binary number that is equivalent to the decimal numbers 0 to 9. Thus the decimal number 9 is equivalent to the binary number 1001

binary number A number which can have just two values, especially the numbers 1 and 0

bipolar transistor A transistor that depends for its operation on both n-type and p-type semiconductors, i.e. its function depends on both electrons and holes

bistable (also **flip-flop**) A circuit that has two outputs which can act as memories for data fed into its input. Flip-flops are used in many types of electronic counter and computer memory

bit (binary digit) Either of the two numbers 0 and 1 which are the basic units of data in computers and other digital systems. A group of bits is known as a word, and a group of 8 bits is known as a byte

Boolean algebra A shorthand way of dealing with logical statements that are true or false. Boolean algebra was developed by George Boole, a 19th century mathematician, and it has since become a useful way of analysing and predicting the behaviour of two-state digital circuits

boule (also **ingot**) A very pure crystal of silicon or other semiconductor from which thin slices (wafers) are cut at the start of the complex process of making silicon chips

byte A group of binary digits, e.g. the byte 10100101, which is handled as a single unit of data in computers and other data handling systems

C The symbol for electrical capacitance

C The symbol in the hexadecimal system for the decimal number 12. Thus the hexadecimal number C6 equals the decimal number $12 \times 16 + 6 = 198$

cable TV (CATV) The distribution of TV programs by means of cables laid underground

CAD (computer-aided design)

CAL (computer-aided learning)

capacitance (C) The ability of a capacitor to store electric charge

capacitor A component that stores electric charge by electrostatic means

carrier wave (CW) A relatively high frequency radio or light wave which carries a message from a transmitter to a receiver

cathode The terminal of a device (e.g. a diode) from which electrons flow

cathode ray oscilloscope (CRO) A test and measurement instrument for showing the patterns of electrical waveforms and for measuring their frequency and other characteristics

CATV (cable TV)

CCD (charge-coupled device)

central processor unit (CPU) The principal operating and controlling part of a computer, also known as its microprocessor

channel The conducting path between the drain and source terminals of a field-effect transistor

charge A basic property of matter which occurs in discrete units, usually equal to the charge on an electron, and which can be of positive or negative polarity

charge-coupled device (CCD) An array of minute semiconductor memory cells on a silicon chip which is used to store an optical image, usually of very faint sources of light. CCD devices are used in security cameras and astronomical telescopes

chip (silicon chip)

clock Any circuit that produces a regular series of on/off pulses (0s and 1s) which are used to synchronise the flow of binary data in digital processing machines such as computers

closed-loop control A method of controlling the output of a system by feeding its input with an error voltage to reduce the difference between the desired and actual outputs

CMOS (complementary metal-oxide semiconductor)

code A set of symbols, e.g. Morse code, or of conventions, e.g. the ASCII code, which represents information in a suitable form for transmission from one place to another

collector One of the three terminals of a bipolar transistor usually connected to the output side of a transistor

combinational logic A digital circuit, e.g. a NAND gate, that produces an output based on the combination of 0s and 1s presented to its inputs

comparator An electronic device, e.g. one based on an operational amplifier, that produces an output when the voltages of two input signals are different

complementary metal-oxide semiconductor (CMOS) A switching circuit based on the combination of n-channel and p-channel field-effect transistors. A wide range of digital devices, e.g. digital watches and microprocessors, are based on CMOS devices in integrated circuit form. These chips have the advantage of low power consumption and they can operate from a wide range of supply voltage

computer A programmable device used for storing, retrieving and processing data

computer-aided design (CAD) A computer system for helping architects, scientists, aircraft designers and others design their products and services

computer-aided learning (CAL) The use of a computer to improve traditional teaching methods using graphics, animation and sound, and having the advantage that it allows a student to progress at his or her own pace

computer generations The family history of computers beginning with the first generation based on valves (1940s), the second generation on transistors (1950s), the third generation on integrated circuits containing several hundred transistors (1960s), the fourth generation on ICs containing several thousand transistors, and the fifth generation based on ICs containing over a million transistors on a single silicon chip

conductivity A measure of how well electricity flows through a substance

coulomb The unit for measuring electrical charge. A current of 1 ampere is produced by electric charge flowing at the rate of 1 coulomb per second

counter Any device, e.g. a decade counter, made from flip-flops and used for counting binary data entering its input

CPU (central processor unit)

CRO (cathode-ray oscilloscope)

cryoelectronics The use of electronics at temperatures close to absolute zero, i.e. near to $-273\,°C$

crystal clock A very stable high frequency oscillator usually based on a crystal of quartz and used, for example, to control and synchronise the various operations inside a computer

cycle A complete sequence of a wave pattern that is repeated at regular intervals

D The symbol used in the hexadecimal system for the decimal number 13. Thus the hexadecimal number 7D equals the decimal number $7 \times 16 + 13 = 125$

Darlington pair An arrangement of two bipolar transistors connected in series so that their combined current gain equals (theoretically) the product of their individual current gains. The Darlington pair is widely used in switching circuits and computer interfacing circuits

debouncer A circuit, usually based on a Schmitt trigger, that ignores unwanted secondary pulses generated by contact bounce of a mechanical switch such as a reed switch

decade counter A binary counter that counts up to a maximum count of ten before resetting to zero

decibel (dB) A unit used for comparing the strengths of two signals, such as the intensity of sound and the voltage gain of an amplifier. The decibel is defined logarithmically so that a doubling of the signal strength is reckoned as an increase of 3 dB

decoder A device that converts coded information, e.g. the binary code, into a more readily understood code such as decimal

demodulator A device for recovering information, e.g. music, from a carrier wave, e.g. a laser beam

demultiplexer A device for recovering the individual messages from a communications channel that carries several messages simultaneously

depletion layer The region across a reverse-biased pn junction that contains few free electrons and holes and which is responsible for the rectifying properties of a diode

detector A device in the demodulator stage of a radio receiver that helps to recover the original signal from the carrier wave

dielectric The insulating layer of, say, mica, glass, or polystyrene, between the conducting plates of a capacitor

diffusion The movement of electrons and/or holes from a region of high to low concentration of these charge carriers

digital computer A system that uses gates, flip-flops, counters, etc. to process information in digital form

digital-to-analogue converter (DAC) A device that converts a digital signal into an equivalent analogue signal. DACs are widely used in computer systems for controlling the speed of motors, the brightness of lamps, etc.

d.i.l. (dual-in-line)

doping The process of introducing minute amounts of material, the dopant, into a silicon to produce n-type or p-type semiconductors in the making of transistors, integrated circuits and other devices

dot matrix display A method of producing clearly seen symbols by selectively illuminating a matrix of LEDs

drain One of the three terminals of a field-effect transistor

driver Any device, e.g. a Darlington pair, that provides sufficient power to operate a load, e.g. a relay

E The symbol used in the hexadecimal system for the decimal number 14. Thus the hexadecimal number E3 equals the decimal number $14 \times 16 + 3 = 227$

electrode A metal connector used to make electrical contact with a circuit

electrolytic capacitor A capacitor which is made from two metal plates separated by a very thin layer of aluminium oxide. Electrolytic capacitors offer a high capacitance in a small volume, but they are polarised and need connecting the right way round in a circuit

electromagnetic spectrum The family of radiations which all travel at the speed of light through a vacuum, and extend from very short wavelength gamma rays to very long wavelength radio waves, and include light, infrared and ultraviolet radiation

electromotive force (e.m.f.) The electrical force generated by a cell or battery that makes electrons move through a circuit connected across the terminals of the battery

electron A small negatively-charged particle which is one of the basic building blocks of all substances and forms a cloud round the nucleus of an atom

electronics The study and application of the behaviour and effects of electrons in transistors, integrated circuits and other devices

electron optics A branch of electronics that deals with the theory and practice of focusing and deflecting electron beams moving through electric and magnetic fields as in a TV receiver

e.m.f. (electromotive force)

emitter One of the three terminals of a bipolar transistor which is usually connected to both the input and output circuits

encoder Any device that converts information into a form suitable for transmission by electronic means

energy level One of the allowed energies that an electron can have in the electron cloud of an atom

epitaxial layer One of a number of thin layers of semiconductor that is formed on a layer of silicon in the process of making an integrated circuit by masking and etching

etching The process of removing silicon dioxide from minute areas of a silicon chip during the several stages involved in the making of an integrated circuit. Following etching, the underlying layer of silicon is doped with an impurity in order to change its electrical properties to an n-type or p-type semiconductor

exclusive OR gate A building block in digital logic circuits that makes the following logical decision: it only produces an output of logic 1 when any one, but not all, of its inputs are at logic 1

F The symbol used in the hexadecimal system for the decimal number 15. Thus the hexadecimal number FF equals the decimal number $15 \times 16 + 15 = 255$

fan-in The number of logic gate outputs which can be connected to the input of another logic gate

farad (F) The unit of electrical capacitance and equal to the charge stored in coulombs in a capacitor when the potential difference across its terminals is 1V

feedback The sending back to the input of part of the output of a system in order to improve the performance of the system. There are two types of feedback, positive and negative

FET (field-effect transistor)

fibre optics The use of hair-thin transparent glass fibres to transmit information on a light beam that passes through the fibre by repeated internal reflections from the walls of the fibre

field-effect transistor (FET) A transistor that depends for its operation on either n-type or p-type semiconductor material. The FET is a unipolar device since its function depends either on the flow of electrons, an n-type FET, or holes, a p-type FET

filter A device for controlling the range of frequencies that passes through a circuit. Thus a low-pass filter in a hi-fi system reduces tape hiss by reducing high frequencies of the audio signal

flip-flop (bistable)

floppy disk A flexible disk, usually 5.25 inches (133 mm) in diameter, made of plastic and coated with a magnetic film on which computer data can be stored and erased

FM (frequency modulation)

forward bias A voltage applied across a pn junction which causes electrons to flow across the junction

frequency The number of times per second that a regular process repeats itself. Frequency is measured in hertz (Hz) or cycles per second. The frequency of the mains supply is 50 Hz

frequency modulation (FM) A method of sending information by varying the frequency of a radio or light wave in response to the amplitude of the message being sent. For high quality radio broadcasts, FM is preferable to AM since it is affected less by interference from electrical machinery and lightning

full-wave rectifier A semiconductor device based on four diodes that produces direct current from alternating current by reversing the flow of current in one half cycle of the alternating current

fuse A device that acts as the 'weak link' in a circuit which it protects from excessive current flow

gain The increase in the power, voltage or current of a signal as it passes through an amplifier

gallium arsenide (GaAs) A crystalline material which, like silicon, is used to make diodes, transistors and integrated circuits. GaAs conducts electricity about seven times faster than silicon, and it is favoured for computer memory devices used in guided missiles and other fast-acting weapons

gamma rays Electromagnetic radiation which has a much shorter wavelength than light. Gamma rays are generated by radioactive substances and are present in cosmic rays

gate One of the three terminals of a field-effect transistor.

gate (logic gate)

germanium A nonmetallic element of atomic number 32 and the first material to be used as the basis of transistors and diodes

gigabyte (GB) A quantity of computer data equal to one thousand million bytes

gigahertz (GHz) A frequency equal to one thousand million hertz (10^9 Hz)

H The symbol for electrical inductance measured in henries

half-wave rectifier A diode, or circuit based on one or more diodes, which produces a direct current from alternating current by removing one half of the AC waveform

hardware Any mechanical or electronic equipment that makes up a system

heat sink A relatively large piece of metal that is placed in contact with a transistor or other component to help dissipate the heat generated within the component

henry (H) The unit of electrical inductance defined as the potential difference generated across the terminals of an inductor when the current through it is changing at the rate of one ampere per second

hertz (Hz) The unit of frequency equal to the number of complete cycles per second of an alternating waveform

hexadecimal number A number which has a base of 16 and is expressed using the numbers 0 to 9 and the letters A to F. Thus the hexadecimal number C7 is equivalent to the decimal number $12 \times 16 + 7 = 199$

hole A vacancy in the crystal structure of a semiconductor that is able to attract an electron. A p-type semiconductor contains an excess of holes which act as mobile charge carriers and move through the semiconductor under the action of an electric field

I The symbol for electric current and measured in amperes
IC (integrated circuit)
impedance (Z) The resistance of a circuit to alternating current flowing through it, and which varies with the frequency of the current
impurity An element such as boron that is added to silicon to produce a semiconductor with desirable electrical qualities
inductance (L) The property of a circuit, especially a coil of wire, that makes it generate a voltage when a current, either in the circuit itself or in a nearby circuit, changes
inductor An electrical component, usually in the form of a coil of wire, that is designed to resist changes to the flow of current through it. Thus inductors are used as 'chokes' to reduce the possibly damaging effects of sudden surges of current, and in tuned circuits
information technology (IT) The gathering, processing and circulation of information by combining the data-processing power of the computer with the message-sending capability of communications
infrared Radiation having wavelengths between the visible red and microwave regions of the electromagnetic spectrum, i.e. wavelengths between about 700 nanometres (7×10^{-9} m) and 1 mm. Though invisible to the naked eye, infrared is widely used in electronics, especially in remote control and security systems
input The point at which information enters a device
input/output port (I/O port) The electrical 'window' on most computer systems that allows the computer to send data to and receive data from an external device
insulator A material, e.g. glass, that does not allow electricity to pass through it
integrated circuit (IC) An often very complex electronic circuit which has resistors, transistors, capacitors and other components formed on a single silicon chip
interface A circuit or device, e.g. a modem, that enables a computer to transfer data to and/or from its surroundings or between computers
ion An atom or group of atoms that has gained or lost one or more electrons and which therefore carries a positive charge
IT (information technology)

joule The SI unit of energy and defined as the work done when a force of 1 newton (1 N) moves its point of application a distance of 1 metre
junction A region of contact between two dissimilar metals (as in a thermocouple) or two dissimilar conductors (as in a diode) which has useful electrical properties

k The symbol used for the prefix 'kilo' meaning one thousand times
K The symbol for a memory capacity equal to 1024 bytes
kilobit One thousand bits, i.e. 0s and 1s, of data
kilobyte One thousand bytes of data
kilohertz (kHz) A frequency equal to 1000 Hz

large-scale integration (LSI) The process of making integrated circuits with between 100 and 5000 logic gates on a single silicon chip

laser A device that produces an intense and narrow beam of light of almost one particular wavelength. The light from lasers is used in optical communications systems, compact disk players and video disk players

LCD (liquid crystal display)

LDR (light-dependent resistor)

least significant bit (LSB) The right-most digit in a binary word, e.g. 1 is the LSB in the word 0101

LED (light-emitting diode)

light-dependent resistor (LDR) A semiconductor device that has a resistance decreasing sharply with increasing light intensity. The LDR is used in light control and measurement systems, e.g. automatic street lights and cameras

light-emitting diode (LED) A small semiconductor diode that emits light when current passes between its anode and cathode terminals. Red, green, yellow and blue LEDs are used in all types of display systems, e.g. hi-fi amplifiers

liquid crystal display (LCD) A display that operates by controlling the reflected light from special liquid crystals, rather than by emitting light as in the light-emitting diode

load The general name for a device, e.g. an electric motor, that absorbs electrical energy to produce mechanical or some other form of energy

logic circuit An electronic circuit that carries out simple logic functions, e.g. a NAND gate

logic diagram A circuit diagram showing how logic gates and other digital devices are connected together to produce a working circuit or system

logic gate A digital device, e.g. an AND gate, that produces an output of logic 1 or 0 depending on the combination of 1s and 0s at its inputs

loudspeaker A device used to convert electrical energy into sound energy. It usually comprises a coil of wire attached to a paper cone located in a strong magnetic field. The coil and cone move when current flows through the coil

LSB (least significant bit)

LSI (large-scale integration)

m The symbol for the prefix 'milli' meaning one thousandth of

M The symbol for the prefix 'mega' meaning one million times

machine code Instructions in the form of patterns of binary digits which enable a computer to carry out calculations

magnetic bubble memory (MBM) A device that stores data as a string of magnetic 'bubbles' in a thin film of magnetic material. The MBM can store a very large amount of data in small volume and is ideal for portable computer products such as word processors

magnetic storage Magnetic tapes, floppy disks and magnetic bubble memories that store data as local changes in the strength of a magnetic field, and which can be recovered electrically

man-machine interface Any hardware, e.g. a keyboard or mouse, that allows a person to exchange information with a computer or machine

mark-to-space ratio The ratio of the time that the waveform of a rectangular waveform is HIGH to the time it is LOW

maser (microwave amplification by stimulated emission of radiation) A device for amplifying very weak radio signals, e.g. those received from interplanetary spacecraft. A maser increases the strength of a radio signal by energising atoms to a level where they give off radio waves at the desired frequency

MBM (magnetic bubble memory)

medium-scale integration (MSI) The process of making integrated circuits with between 120 and 1000 logic gates on a single silicon chip

medium waves Radio waves having wavelengths in the range about 200 to 700 m, i.e. frequencies in the range 1.5 to 4.5 MHz

megabit A quantity of data equivalent to one million (10^6) bits

megabyte A quantity of binary data equal to one million (10^6) bytes. Present day floppy disks store about this amount of data

megahertz (MHz) A frequency equal to one million (10^6) Hz. The crystal clock in a microcomputer oscillates at a frequency of a few megahertz

megohm (MΩ) An electrical resistance equal to one million (10^6) ohms

memory That part of a computer system used for storing data until it is needed. A microprocessor in a computer can locate and read each item of data by using an address

microcomputer A usually portable computer which can be programmed to perform a large number of functions quickly and relatively cheaply. Its main uses are in the home, school, laboratory and office for playing computer games, helping with studies, planning home and business management, controlling domestic heating systems, and so on

microelectronics The production and use of complex circuits on silicon chips

microfarad A unit of electrical capacitance equal to one millionth of a farad

micron (micrometre) A distance equal to one millionth of a metre. The micron is used for measuring the size and separation of components on silicon chips

microphone A device that converts sound waves into electrical signals usually for subsequent amplification

microprocessor A complex integrated circuit manufactured on a single silicon chip. It is the 'heart' of a computer and can be programmed to perform a wide range of functions. A microprocessor is used in washing machines, cars, cookers, games and many other products

microsecond (μs) A time interval equal to one millionth (10^{-6}) of a second

microswitch A small mechanically-operated switch that usually needs only a small force to operate it

microwaves Radio waves having wavelengths less than about 300 mm and used for straight line communications by British Telecom and others

millimetre (mm) A distance equal to one thousandth (10^{-3}) of a metre

millisecond (ms) A time equal to one thousandth (10^{-3}) of a second

milliwatt (mW) A power equal to one thousandth (10^{-3}) of a watt

minority carrier The least abundant of the two charge carriers present in a semiconductor. The minority charge carriers in n-type material are holes

modem (modulator/demodulator) A device for converting computer data in digital form into analogue signals for transmission down a telephone line

modulator A circuit that puts a message on some form of carrier wave, e.g. light waves, for transmission in a communications system

monostable A circuit that produces a time delay when it is triggered, and then reverts back to its original, normally stable, state

most significant bit (MSB) The left-most binary digit in a digital word

mouse A small hand-operated device connected to a computer by a trailing wire, or by optical means, that makes a cursor move around the screen of a VDU to select operations and make decisions

MSB (most significant bit)

MSI (medium-scale integration)

multimeter An instrument for measuring current, potential difference and resistance and used for testing and fault-finding in the design and use of electronic circuits

multiplexing A method of making a single communications channel carry several messages

multivibrator Any one of three basic types of two-stage transistor circuit in which the output of each stage is fed back to the input of the other stage using coupling capacitors and resistors, and causing the transistors to switch on and off rapidly. The multivibrator family includes the monostable, astable and bistable

n The symbol for the prefix 'nano' meaning one thousand millionth of

NAND gate A building block in digital logic circuits that makes the following logical decision: it produces an output of logic 1 when one or more of its inputs are at logic 0, and an output of logic 0 when all its inputs are at logic 1

nanofarad A unit of electrical capacitance equal to one thousand millionth (10^{-9}) of a farad

nanosecond A time interval equal to one thousand millionth of a second

n-channel FET A field-effect transistor that is constructed so that electrons flow along an n-type conducting layer (the channel) between its source and drain terminals under the control of a potential difference between its gate and source terminals

negative feedback The feeding back to the input of a system part of its output signal so as to reduce the effect of the input. Negative feedback is widely used in amplifiers and control systems to improve their stability

neutron A particle in the nucleus of an atom which has no electrical charge and a mass roughly equal to that of the proton

NOR gate A building block in digital logic circuits that makes the following logical decision: it produces an output of logic 1 when all its inputs are at logic 0, and an output of logic 0 when one or more of its inputs are at logic 1

NOT gate A building block in digital logic circuits that makes the following logical decision; it produces an output of logic 1 when its single input is at logic 0, and vice versa

npn transistor A semiconductor device that is made from both p-type and n-type semiconductors and which is used in switching and amplifying circuits. An npn transistor has three terminals, emitter, collector and base. The current flowing between the emitter and collector terminals is controlled by a small current flowing into or out of its base terminal

n-type semiconductor A semi-conductor through which current flows mainly as electrons

nucleus The central and relatively small part of an atom that is made up of protons and neutrons

ohm (Ω) The unit of electrical resistance

Ohm's law The potential difference across the ends of a metallic conductor is proportional to the current flowing through it if its temperature, shape and other physical factors remain unchanged

operational amplifier (op. amp.) A very high gain amplifier that produces an output voltage proportional to the difference between its two input voltages. Op. amps are widely used in instrumentation and control systems, e.g. in thermometers and thermostats, and they were used in the now almost obsolete analogue computers

optical communications The use of long thin optical fibres for sending messages using pulses of laser light

optical fibre A thin glass or plastic thread through which light travels without escaping from its surface. Optical fibres have their decorative uses, but their greatest impact is in optical communications

optical memory A method of storing computer data in digital form which is read by optical means. Optical memory devices include video disks and holographic memory

optocoupler An optoelectronic device that sends signals from one part of a circuit to another by means of light beams passing through air. Optocouplers are widely used for electrically isolating two parts of a system, e.g. between a microcomputer and a mains-operated lamp

optoelectronics A branch of electronics dealing with the interaction between light and electricity. Light-emitting diodes and liquid crystal displays are examples of optoelectronic devices

OR gate A building block in digital logic circuits that makes the following logical decision: it produces an output of logic 1 when one or more of its inputs are at logic 1, and an output of logic 0 when all its inputs are at logic 0

oscillator A circuit or device, e.g. an audio frequency oscillator, that provides a sinusoidal or square wave voltage output at a chosen frequency. An astable is one type of oscillator

p The symbol for the prefix 'pico' meaning one million millionth of

package The plastic or ceramic material that is used to cover and protect an integrated circuit

parallel circuit A circuit in which components, e.g. resistors, are connected side-by-side so that current is shared by the components

PCB (printed circuit board)

p-channel FET A field-effect transistor that is constructed so that holes flow through a p-type conducting layer (the channel) between its source and drain terminals under the control of a potential difference between its gate and source terminals

period The time taken for a wave motion, e.g. a radio wave, to make one complete oscillation. The period of the 50 Hz mains frequency is 0.02s

photodiode A light-sensitive diode that has two terminals, a cathode and an anode, and that responds rapidly to changes of light

photoelectron An electron that is released from the surface of a metal by the action of light

photomask A transparent glass plate used in the manufacture of integrated circuits on a silicon chip. A photomask contains a precise pattern of microscopically small opaque 'spots' that has been produced by the photographic reduction of a much larger pattern

photon   The smallest 'packet', or quantum, of light energy

photoresist   A light-sensitive material that is spread over the surface of a silicon wafer from which silicon chips are made, and whose solubility in various chemicals is altered by exposure to ultraviolet light. The photoresist is used with a photomask so that holes can be etched at selected points in the surface of the silicon

photoresistor   A transistor that responds to light and produces an amplified output signal. Like photodiodes, photoresistors respond rapidly to light changes and are used as sensors in optical communications systems

picofarad (pF)   An electrical capacitance equal to one million millionth of a farad (10^{-12}F)

piezoelectricity   The electricity that certain crystals (e.g. quartz) produce when they are squeezed. Conversely, if a potential difference is applied across a piezoelectric crystal, it alters shape slightly. The piezoelectric effect is put to good use in digital watches, hi-fi pick-ups and gas lighters

pnp transistor   A semiconductor device that is made from both p-type and n-type semiconductors and which is used in switching and amplifying circuits. A pnp transistor has three terminals, emitter, collector and base. The current flowing between the emitter and collector terminals is controlled by a small current flowing into or out of the base terminal

port   A place on a microcomputer to which peripherals can be connected to provide two-way communication between the computer and the outside world

positive feedback   The feeding back to the input of a system a part of its output signal so as to increase the effect of its input. Positive feedback is used in an astable

potential divider   Two or more resistors connected in series through which current flows to produce potential differences dependent on the resistor values

potentiometer   An electrical component having three terminals that provides an adjustable potential difference

printed circuit board (PCB)   A thin board made of electrically insulating material (usually glass fibre) on which a network of copper tracks is formed to provide connections between components soldered to the tracks from the other side of the board.

proton   A particle that makes up the nucleus of a hydrogen atom, that coexists with neutrons in the nuclei of all other atoms, and has a positive charge equal in value to the negative charge on an electron

p-type semiconductor   A semiconductor through which current flows mainly as holes

pulse   A short-lived variation of voltage or current in a circuit, or of an electromagnetic wave such as laser light

Q-factor   The sharpness (or 'quality') of an electronic filter circuit, e.g. a tuned circuit, that enables it to accept or reject a particular frequency

quantum   The smallest packet of radiant energy, e.g. a photon, that can be transmitted from place to place. The quantum of energy is proportional to the frequency of the radiation so that X-ray quanta carry more energy than infrared quanta

quartz A crystalline form of silicon dioxide that has two useful properties: it is piezoelectric so that it can be used to provide stable frequencies in crystal clocks; and it is transparent to ultraviolet light so that it can be used as a window to cover an electrically-programmable read-only memory

qwerty keyboard A keyboard (e.g. a computer keyboard) that has its keys arranged in the same way as those of a standard typewriter, i.e. the first six letters of the top row spell 'QWERTY'

R The symbol for electrical resistance, the ohm

radar (**ra**dio **d**etection **a**nd **r**ange-finding) An electronic system for locating the position of distant objects by recording the echo of high frequency radio waves bounced off the objects

radio Communication at a distance by means of electromagnetic waves having frequencies in the range about 15 kHz to 100 MHz

RAM (random-access memory)

random access memory (RAM) An integrated circuit that is used for the temporary storage of computer programs

read-only memory (ROM) An integrated circuit that is used for holding data permanently, e.g. for storing the language and graphics symbols used by a computer

rectifier A semiconductor diode that makes use of the one-way conducting properties of a pn junction to convert a.c. to d.c.

relay A magnetically operated switch that enables a small current to control a much larger current in a separate circuit

resistance The opposition offered by a component to the passage of electricity through it

resistor A component that offers resistance to electrical current

resonance The build-up of large amplitude oscillations in a circuit, e.g. a tuned circuit, by feeding it with alternating current close to the natural frequency of oscillation of the circuit

reverse bias A voltage applied across a pn junction (e.g. a diode) which prevents the flow of electrons across the junction

RGB monitor A VDU that displays computer graphics and text in colour using the red, green and blue electron guns in the monitor

robot A computer-controlled device that can be programmed to perform repetitive tasks such as paint-spraying, welding, and machining of parts

ROM (read-only memory)

root mean square (r.m.s.) value The steady d.c. current that has the same heating effect as a sinusoidal alternating current. Thus 240 V is the r.m.s. value of the mains voltage

s The symbol for a second of time

S The symbol for the unit of electrical conductance, the siemen

Schmitt trigger A snap-action electronic switch which turns off and on at two specific input voltages. The Schmitt trigger is widely used to 'sharpen up' slowly changing waveforms, and to eliminate noise from signals input to circuits such as thermostats and counters

semiconductor A solid material that is a better electrical conductor than an insulator (e.g. polythene), but not such a good conductor as a metal (e.g. silver). Diodes, transistors and integrated circuits are based on n-type and p-type semiconductors

sensor Any device (e.g. a thermistor) that produces an electrical signal indicating a change (e.g. temperature) in its surroundings

sequential logic A digital circuit that can store information. Sequential logic circuits are based on flip-flops and are the basis of counters and computer memories

servosystem An electromechanical system which uses sensors to precisely control and monitor the movement of something (e.g. the read/write head of a floppy disk drive)

short waves Radio waves that have wavelengths between about 2.5 MHz and 15 MHz, and which are mainly used for amateur and long-range communications

siemen The unit of electrical conductance equal to the reciprocal ohm. Thus as resistance increases, conductance decreases

silicon An abundant nonmetallic element that has largely replaced germanium for making diodes and transistors. Silicon is doped with small amounts of impurities such as boron and phosphorus to make n-type and p-type semiconductors

silicon chip A small piece of silicon about the size of this letter 'O' on which a complex miniaturised circuit (called an integrated circuit) is formed by photographic and chemical processes

small-scale integration (SSI) The process of making integrated circuits comprising less than 20 logic gates per silicon chip

software Instructions or programs stored in a computer system (e.g. on disk or read-only memory) and which is manipulated by the system

solder An alloy of tin and lead (roughly in the ratio 60:40) that has a low melting point and is used for making electrical connections between components on a circuit board

solenoid A coil of copper wire in which an iron rod moves by the magnetic field produced when a current flows through the coil

SSI (small-scale integration)

stepping motor An electric motor with a shaft that rotates one step at a time taking, for example, 48 steps to complete one revolution. Stepping (or stepper) motors are used for the precise positioning of robot arms and printer paper, for example, under computer control

strain gauge A sensor attached to an object to detect how much it lengthens or shortens when it is loaded. The change in electrical resistance of the strain gauge is a measure of the distortion produced

system All parts making up a working whole

tape A long strip of plastic material with a magnetisable surface used to record information

telecommunications The use of electronic and other equipment to send information through wires, the air and interplanetary space

telemetry The transmission by radio of measurements made at a distance, e.g. from automatic weather stations or interplanetary space probes

television (TV) The transmission and reception of pictures using high frequency radio waves

thermal runaway Damage to an electronic device (e.g. a transistor) caused by the progressive rise in current through it as its temperature increases

thermistor A device made from a mixture of semiconductors such that its resistance changes with temperature. Thermistors are widely used in thermostats and thermometers

thermocouple A device made from a pair of dissimilar metals (e.g. copper and iron) that produces a voltage varying with temperature. Thermocouples are ideal as sensors in electronic thermometers since the junction between the metals can be made very small

time constant The time taken for the voltage across a capacitor to rise to 63% of its final voltage when it charges through a resistor connected in series with it

transducer A device (e.g. a thermocouple) that changes one form of energy into another. Transducers are widely used in electronic sensing and control systems

transformer An electromagnetic device for converting alternating current from one voltage to another

transistor A semiconductor device (e.g. a bipolar transistor) that has three terminals and is used for switching and amplification

transistor-transistor logic (TTL) A type of digital IC, based on bipolar transistors, that provides logic and counting functions and requires a 5 V supply

truth table A list of 0s and 1s that shows how a digital logic circuit (e.g. a NAND gate) responds to all possible combinations of binary input signals

TTL (transistor-transistor logic)

tuned circuit A circuit that contains an inductor and a capacitor and can be tuned to receive particular radio signals

UHF (ultra-high frequency)

ultra-high frequency (UHF) Radio waves that have frequencies in the range 500 MHz to 30000 MHz and are used for TV broadcasts

ultrasonic waves Sound waves inaudible to the human ear that have frequencies above about 20 kHz. Ultrasonic waves are detected and generated by transducer devices used in remote control devices

ultraviolet Radiation having wavelengths between the visible violet and the X-ray region of the electromagnetic spectrum, i.e. wavelengths between 400 nanometres (nm) and 2 nm. Ultraviolet radiation is used in the manufacture of integrated circuits since it produces chemical changes to a photoresist in the manufacture of integrated circuits and printed circuit boards. It is also used for erasing data from erasable programmable read-only memories (EPROMs)

unipolar transistor A transistor that depends for its operation on either n-type or p-type semiconductor materials as in a field-effect transistor

VDU (video display unit)

VLSI (very large scale integration)

very large scale integration (VLSI) The process of making integrated circuits with in excess of 5000 logic gates on a silicon chip

very high frequency (VHF) Radio waves that have frequencies in the range 30 MHz to 300 MHz and that are used for high quality radio broadcasts (FM) and TV transmission

VHF (very high frequency)

video camera A device for recording sound and vision on magnetic video tape for playback on a TV set. Today's video cameras are becoming lighter and smaller and now make use of charge-coupled devices (CCDs) to detect the image and convert it into electrical signals

video disk A disk on which picture and sound information are recorded in digital form for playback in a video disk player. One type of video disk player uses a finely focused laser beam to detect the microscopic pits on a continuous spiral track on the surface of the rapidly rotating disk. Video disks do not allow programmes to be recorded from the TV receiver

video display unit (VDU) An input/output device comprising a screen and sometimes a keyboard that enables a person to communicate with a computer

video tape A magnetic tape on which programmes are recorded in sound and vision, and which is held in a video cassette for playing on a video cassette recorder

video text Any system that uses VDUs to display computer-based information to people at home and work. There are two main types of videotext: broadcast videotext (known as teletext); and interactive videotext (known as viewdata)

viewdata An information service that enables telephone subscribers to access a wide range of information held in a database, and which is displayed on a TV set coupled to the telephone line by a modem

VMOS (vertical metal-oxide semiconductor)

vertical metal-oxide semiconductor (VMOS) A type of field-effect transistor that is constructed on a silicon chip by etching a V-shaped groove deep inside the chip. This process provides a small-size, high-power and fast-acting transistor for use in audio amplifiers and power switching circuits

volt (V) The unit of electrical potential difference that causes a current of one ampere to flow through a resistance of one ohm

voltage The number of volts at a point in a circuit relative to the circuit's zero potential

voltmeter An instrument for measuring the electrical potential in volts between two points in a circuit

W The symbol for the watt, the unit of power

wafer A thin disk cut from a single crystal of silicon on which hundreds of integrated circuits are made before being cut up into individual ICs for packaging

watt (W) The unit of power and equal to the rate of conversion of energy of 1 joule per second

waveform The shape of an electrical signal, e.g. a sinusoidal waveform

wavelength The distance between one point on a wave and the next corresponding point (e.g. from crest to crest). Wavelength is related to the frequency and the speed of the wave by the simple equation, speed = wavelength × frequency. The speed of radio waves and light waves in a vacuum is about 3×10^8 metres per second

Wheatstone bridge A device consisting of four resistors connected in series and providing a resistance to voltage conversion in some types of electronic instrument

word A pattern of bits (i.e. 1s and 0s) that is handled as a single unit of information in digital systems; e.g. a byte is an 8-bit word

wordprocessor A computerised typewriter that allows written material to be generated, stored, edited, printed and transmitted. A wordprocessor comprises a keyboard (and/or a mouse) for inputting text, a microprocessor for making decisions about the text, a VDU for seeing what is being written, a floppy disk drive for storing text, and a printer for hard copy

X-rays Penetrating electromagnetic radiation used in industry and in medicine for seeing below the surface of solid materials

zener diode A special semiconductor diode that is designed to conduct current in the reverse-bias direction at a particular reverse-bias voltage. Zener diodes are widely used to provide stabilised voltages in electronic circuits

Index

A selection of further titles
from Hodder & Stoughton / Teach Yourself

0 340 40713 1   Electricity   *David Bryant*   £5.99   ☐

All Hodder & Stoughton / Teach Yourself books are available from your local bookshop or can be ordered direct from the publisher. Just tick the titles you want and fill in the form below. Prices and availability subject to change without notice.

To: Hodder & Stoughton Ltd, Cash Sales Department, Bookpoint, 73C Milton Park, Abingdon, OXON, OX14 4TD, UK. If you have a credit card you may order by telephone – 01235 831700.

Please enclose a cheque or postal order made payable to Bookpoint Ltd to the value of the cover price and allow the following for postage and packing:

UK & BFPO: £1.00 for the first book, 50p for the second book and 30p for each additional book ordered up to a maximum charge of £3.00.
OVERSEAS & EIRE : £2.00 for the first book, £1.00 for the second book and 50p for each additional book.

Name ....

Address ....

....

....

If you would prefer to pay by credit card, please complete:
Please debit my Visa / Access / Diner's Card / American Express (delete as appropriate)
card no:

Signature .... Expiry Date ....